45 Presidents through the eyes of AI

By

Elizabeth Bode Firnhaber

Copyright © 2025 by – Elizabeth Bode Firnhaber – All Rights Reserved.

No part of this book may be reproduced or transmitted in any form by any means, whether graphic, electronic, or mechanical, including photography, recording, taping, or by any information storage or retrieval system without prior written permission from the author.

The Library of Congress Control Number:
2025947955

ISBN:
978-1-966903-93-2

Dedication

To my Lord, Jesus, my immediate family: Paul, Jean, Jim, John Jillian, Laura, Josh, John, and my nieces and nephews.

Acknowledgements

Crafting this book required navigating the unfamiliar terrain of modern technology, a challenge for my 94-year-old perspective. Learning to access Grok, compose precise AI prompts to generate content on specific presidential topics, and transfer those words into a document proved a daunting yet exhilarating endeavor. Equally novel was the task of sourcing and selecting images to grace the opening of each chapter—a process I had never undertaken before. My deepest gratitude goes to my son, Jim, whose patient guidance introduced me to these digital tools and ignited this creative journey. I am equally thankful for my daughter, Jean, who, living next door, answered my countless calls for assistance with unwavering support. I also extend my appreciation to Mark Cowan, my skilled technical support, whose expertise resolved computer issues that threatened to derail this project. Without their collective encouragement and expertise, this book would not have come to fruition.

About the Author

Elizabeth was born into a farming family in Courtland, Minnesota, where she grew up as the second eldest of four siblings. Her early life was shaped by strong values, hard work, and a deep connection to community. She graduated from Dr. Martin Luther High School in New Ulm, Minnesota, before earning a bachelor's degree in Communications and Religious Studies from the University of Nebraska, Omaha. She later went on to complete a master's degree in Counseling at Liberty University in Virginia.

Elizabeth's relationship with language began long before she ever picked up a pen. Until the age of five, she spoke only German. Entering first grade without knowing English brought a quiet storm of confusion and frustration—an experience that would deeply shape her sensitivity to language, learning, and the power of communication. In a full-circle moment years later, she became a graduate teaching assistant at the University of Nebraska. In addition to being a graduate teaching assistant, she also taught English to foreign students through a separate program. She knew what it felt like to be on the other side of understanding, and she taught with patience and heart.

In the wake of World War II and amid a national teacher shortage, Elizabeth enrolled in a six-week intensive training program at Concordia College in St. Paul, Minnesota. Designed to prepare students to teach in Lutheran schools, the program also required ongoing study toward a degree. At just seventeen years old, Elizabeth began her teaching career, instructing a seventh-grade class that included a fifteen-year-old student.

Elizabeth was married for fifty-six years until the passing of her beloved husband. Together, they raised four children—Paul, a chiropractor; Jean, a hand therapist; Jim, an airline pilot; and John, a real estate broker. She is also a proud grandmother to two granddaughters.

A woman of many interests, Elizabeth has embraced a lifelong love for reading, teaching Bible classes, singing in choirs and as a soloist, painting, gardening, and playing cards. Now, at the age of ninety-four, she resides in a retirement community in Murrieta, California, where she remains actively engaged in many of her passions.

This book marks her second published work and stands as a testament to her enduring love of language, her commitment to learning, and her desire to connect with others through thoughtful, empathetic storytelling. Elizabeth's journey is proof that even the most challenging beginnings can give rise to voices that inspire, uplift, and endure.

Table of Contents

Dedication ii

Acknowledgements iii

About the Author iv

Prologue 1

George Washington: Father of the Country 3

John Adams: Architect of American Independence 13

Thomas Jefferson: Architect of American Democracy 24

James Madison: Father of the Constitution 36

James Monroe: Architect of the Monroe Doctrine 43

John Quincy Adams: Architect of American Diplomacy 52

Andrew Jackson: The Hero of New Orleans 62

Martin Van Buren: Architect of the Democratic Party 73

William Henry Harrison: The Briefest Presidency 81

John Tyler: The Accidental President 90

James K. Polk: Architect of American Expansion 102

Zachary Taylor: Hero of the Mexican-American War 110

Millard Fillmore: The Compromise President 124

Franklin Pierce: Architect of the Kansas-Nebraska Act 133

James Buchanan: The President Who Failed to Avert Civil War 141

Abraham Lincoln: The Great Emancipator 150

Andrew Johnson: The Reconstruction President 159

Ulysses S. Grant: Hero of the Civil War and Reconstruction 167

Rutherford B. Hayes: Advocate for Civil Service Reform 177

James A. Garfield: The Assassinated Reformer 187

Chester A. Arthur: The Unexpected Reformer 198

Grover Cleveland: The First President to Serve Two Non-Consecutive Terms 207

Benjamin Harrison: The Centennial President 218

Grover Cleveland: (2nd Term) The First President to Serve Non-Consecutive Terms 226

William McKinley: Champion of American Prosperity and Expansion 237

Theodore Roosevelt: Champion of the Progressive Era 248

William Howard Taft: The Trust-Busting President 258

Woodrow Wilson: Architect of the League of Nations 268

Warren G. Harding: The Teapot Dome Scandal President 277

Calvin Coolidge: The Silent Leader of Prosperity 286

Herbert Hoover: The Great Humanitarian 296

Franklin Delano Roosevelt 306

Harry S. Truman: The Man from Missouri 317

Dwight D. Eisenhower: Supreme Allied Commander and Cold War Leader 326

John F. Kennedy: The Camelot President 337

Lyndon B. Johnson: Architect of the Great Society 346

Richard Nixon: The Watergate Scandal President 357

Gerald Ford: The Unelected President 367

Jimmy Carter: Champion of Peace and Humanitarianism ... 376

Ronald Reagan: The Great Communicator 388

Bill Clinton: The Comeback Kid 407

George W. Bush: Architect of the Post- 9/11 Era................. 416

Barack Obama: The First African American President 427

Donald John Trump: Architect of MAGA and Polarizing Political Figure... 438

Joe Biden: Champion of Resilience and Unity..................... 450

Donald J. Trump: The Populist President 462

Addendum.. **475**

Prologue

The story of America's presidency is a tapestry woven from the lives of forty-five men, each marked by distinct virtues, flaws, and ambitions. These leaders, spanning centuries, have carried noble aspirations into the White House, yet none have been immune to human imperfection. Some arrived shadowed by personal failings - intemperance, infidelity, prejudice, or self-interest—while others bore eccentricities that endeared them to some and alienated others. Their humanity, in all its complexity, shaped their tenures and, in turn, the nation. Many presidents endured profound personal loss, particularly in the fledgling years of the republic. They buried children claimed by diseases now conquered and mourned parents whose absence left orphans in their care, some of whom they raised as their own. These stories of resilience and sorrow, too often forgotten, deserve to be shared with new generations, lest we lose sight of the sacrifices that forged our history.

In crafting this book, I have enlisted artificial intelligence to illuminate the multifaceted lives of these leaders, revealing their humanity beyond the marble busts and history's judgment. Each chapter explores a president's early years, family bonds, children, and path to power, alongside the influences that shaped their worldview. Their political affiliations, significant achievements, and the vital contributions of their First Ladies are examined, as are their religious convictions, cherished family pets, and the personality traits—both admirable and flawed—that defined their leadership. Interwoven are compelling anecdotes, details of their passing, the ages at which they and their First Ladies died, and their final resting places.

My hope is that this book will not only deepen your understanding of these forty-five men but also offer a clearer lens through which to view the forces that have shaped our nation's journey. Their triumphs and trials, etched into the fabric of America, remind us that leadership is as much a human endeavor as it is a public one.

George Washington: Father of the Country

George Washington, the first President of the United States, is widely regarded as a foundational figure in American history. Known for his leadership during the American Revolutionary War and his role in shaping the new nation's government, Washington set precedents that defined the presidency. His commitment to national unity and his decision to step down after two terms established a model of democratic leadership.

Introduction

George Washington's life encapsulates the birth of the United States as a nation. Born into a Virginia planter family, he rose from a surveyor and soldier to a revered commander and statesman. His leadership in the Revolutionary War and his presidency laid the groundwork for American governance. Often called the "Father of His Country," Washington's legacy endures through his contributions to independence, governance, and national identity.

Early Life

Washington was born on February 22, 1732, in Westmoreland County, Virginia. His early years were shaped by the rural plantation life of the Virginia gentry. He received a modest education, focusing on practical skills like mathematics and surveying. The death of his father when Washington was 11 forced him to take on responsibilities early, shaping his disciplined and self-reliant character. As a teenager, he worked as a surveyor, gaining knowledge of land and frontier life that later informed his military strategies.

Family

Washington's family ties were rooted in Virginia's colonial elite. His father, Augustine Washington, was a planter and landowner, and his mother, Mary Ball Washington, was known for her strong-willed nature. Washington was the eldest of six children from his father's second marriage. He maintained close ties with his half-brother Lawrence, whose death in 1752 profoundly influenced Washington's early career and inheritance of the Mount Vernon estate.

Children

George Washington had no biological children with his wife, Martha, likely due to health issues from his youth. However, he was a devoted stepfather to Martha's children from her previous marriage.

- **John Parke Custis (Jacky)**: Martha's son, whom Washington guided but struggled with due to Jacky's lack of discipline. He died of camp fever during the Revolutionary War in 1781.

- **Martha Parke Custis (Patsy)**: Martha's daughter, who suffered from epilepsy and died at 17 in 1773. Washington was deeply affected by her death.
- Washington also raised two of John's children, Eleanor Parke Custis and George Washington Parke Custis, after John's death, treating them as his own.

Rise to Power

Washington's ascent to prominence began with his military service and land ownership. His surveying work gave him insight into Virginia's frontier, and his half-brother Lawrence's connections introduced him to influential figures. Key milestones include:

- **French and Indian War (1754–1763)**: Washington gained fame as a young officer, particularly after surviving the disastrous Battle of Monongahela.
- **Virginia House of Burgesses**: Elected in 1758, he honed his political skills.
- **Commander-in-Chief**: Appointed in 1775 by the Continental Congress to lead the Continental Army, his leadership during the Revolutionary War cemented his status as a national hero.
- **Constitutional Convention (1787)**: Washington presided over the convention, lending credibility to the new Constitution and positioning himself as a unifying figure.

Influences

Washington was shaped by a blend of practical experience and Enlightenment ideals. Key influences include:

- His half-brother Lawrence, who introduced him to military life and Virginia's elite.
- The Enlightenment thinkers, whose ideas of liberty and governance informed his belief in a strong but limited government.
- His experiences in the French and Indian War taught him resilience and strategy.
- Virginia's planter class instilled values of duty, honor, and land stewardship.

Party Affiliation

Washington was famously nonpartisan, wary of factionalism that could divide the young nation. He did not formally align with any political party, though his policies often leaned toward Federalist principles, favoring a strong central government, economic stability, and neutrality in foreign affairs. His aversion to party politics set a precedent for presidential impartiality, though tensions between Federalists and Democratic-Republicans emerged during his administration.

Presidency

Washington served as the first U.S. President from April 30, 1789, to March 4, 1797, for two terms. His presidency established critical precedents for the executive branch and American governance.

- **Accomplishments**:
  - Established the executive branch's structure, including the creation of the Cabinet (e.g., appointing Thomas

Jefferson as Secretary of State and Alexander Hamilton as Secretary of the Treasury).

- Signed the Bill of Rights into law, ensuring individual liberties.
- Oversaw the establishment of a national bank, stabilizing the economy.
- Maintained neutrality in foreign conflicts, notably through the Proclamation of Neutrality (1793) during the French Revolutionary Wars.
- Negotiated the Jay Treaty (1794) with Britain, averting war and securing trade.
- Quelled the Whiskey Rebellion (1794), affirming federal authority.
- Set the precedent of a peaceful transition of power by voluntarily stepping down after two terms.
- Delivered the Farewell Address, warning against partisan divisions and foreign entanglements.

First Lady's Contributions

Martha Washington, as the First Lady, defined the role with grace and hospitality, setting a standard for future presidential spouses.

- Hosted social events at the presidential residences in New York and Philadelphia, fostering unity among political leaders.

- Supported Revolutionary War efforts by visiting Washington's camps, sewing clothes, and boosting soldiers' morale.
- Managed Mount Vernon's operations during Washington's absences, demonstrating business acumen.
- Acted as a stabilizing influence on Washington, providing emotional support during his presidency.

Positive Traits and Effects

Washington's character profoundly shaped his presidency. His positive traits include:

- **Integrity**: His refusal to overstep constitutional bounds earned public trust, solidifying the presidency's legitimacy.
- **Leadership**: His calm, decisive leadership unified a fractious nation, particularly during crises like the Whiskey Rebellion.
- **Humility**: Voluntarily stepping down after two terms prevented the presidency from becoming monarchical.
- **Pragmatism**: His balanced approach to governance bridged divides between Federalists and Anti-Federalists.

These traits fostered stability and established the presidency as a respected institution, though his deliberative style sometimes slowed decision-making.

Negative Traits and Effects

Washington's flaws, while few, had notable impacts:

- **Aloofness**: His reserved demeanor could distance him from advisors, occasionally hindering communication within his administration.

- **Limited Formal Education**: His lack of deep theoretical knowledge led him to rely heavily on advisors like Hamilton and Jefferson, whose rivalries complicated governance.

- **Ownership of Slaves**: Washington's failure to address slavery publicly, despite private misgivings, left a moral blind spot that conflicted with his advocacy for liberty.

His slaveholding tarnished his legacy, and he avoided confronting a divisive issue, while his aloofness sometimes strained relations with subordinates.

Pets

Washington was an avid animal lover, particularly fond of horses and dogs, which he kept at Mount Vernon.

- **Horses**: His favorite was Nelson, a chestnut charger ridden during the Revolutionary War, who was later retired to Mount Vernon.

- **Dogs**: Washington bred hunting dogs, naming them whimsically (e.g., Sweetlips, Vulcan, and Tipsy). He is credited with helping develop the American Foxhound breed.

- **Other Animals**: Mount Vernon housed livestock, including mules, which Washington promoted for agricultural efficiency.

Religious Persuasion

Washington was a devout Anglican (later Episcopalian), though his faith was private and pragmatic. He attended church regularly, particularly at Christ Church in Alexandria, Virginia, and valued religion's role in promoting morality. However, he avoided overt displays of piety and was tolerant of other faiths, reflecting Enlightenment principles. His writings rarely mention specific Christian doctrines, suggesting a deistic lean, though he never explicitly identified as such.

Interesting Anecdotes

- **Cherry Tree Myth**: The story of young Washington chopping down a cherry tree and confessing, "I cannot tell a lie," was fabricated by biographer Mason Weems but became a cultural symbol of his honesty.

- **Surviving Battle**: During the 1755 Battle of Monongahela, Washington had two horses shot out from under him and four bullet holes in his coat, yet he emerged unscathed, earning a reputation for resilience.

- **False Teeth**: Contrary to myth, Washington's dentures were not wooden but made of ivory, human teeth, and animal bone, reflecting his lifelong dental issues.

- **Farewell Address**: Washington worked closely with Alexander Hamilton to craft his famous Farewell

Address, which remains a cornerstone of American political thought.

Ages at Death, Causes of Death, and Burial Locations

- **George Washington:**
- **Age at Death:** 67
- **Cause of Death:** Died on December 14, 1799, from a throat infection (likely diphtheria or pneumonia) exacerbated by excessive bloodletting, a common but harmful medical practice of the time.
- **Burial Location:** Mount Vernon, Virginia, in a family vault he designed.
- **Martha Washington:**
- **Age at Death:** 70
- **Cause of Death:** Died on May 22, 1802, likely from old age or a fever, though exact details are unclear.
- **Burial Location:** Mount Vernon, Virginia, alongside George in the family vault.

Conclusion

George Washington's life was a cornerstone of American independence and governance. From his early days as a surveyor to his leadership in war and peace, he embodied duty, resilience, and vision. His presidency set enduring precedents, while Martha's contributions as First Lady provided stability and grace. Despite flaws, particularly his complicity in slavery, Washington's legacy as the "Father of His Country" endures for his role in uniting a fledgling nation and establishing a democratic framework that continues to shape the United States.

John Adams: Architect of American Independence

Introduction

John Adams, the second President of the United States, was a pivotal figure in the founding of the nation, known for his fierce commitment to independence and his role in shaping the early republic. A lawyer, diplomat, and statesman, Adams was a key advocate for breaking away from British rule and establishing a government rooted in law and principle. His presidency, though marked by challenges, solidified key aspects of American governance, and his intellectual rigor and dedication to public service left a lasting legacy.

Early Life

John Adams was born on October 30, 1735, in Braintree (now Quincy), Massachusetts, to a modest farming family. His early life was shaped by Puritan values, education, and a strong sense of duty. He developed a love for learning and a sharp intellect that would define his career.

- Grew up in a rural, agricultural setting, helping with farm chores.
- Attended local schools, showing early aptitude for academics.
- Enrolled at Harvard College at age 15, graduating in 1755 with a degree in law.
- Initially considered a career in the ministry but chose law, believing it better suited to his analytical mind.

Family

Adams married Abigail Smith in 1764, forming a partnership that was both personal and intellectual. Abigail, known for her wit and wisdom, was a crucial influence on Adams, serving as his confidante and advisor. Their marriage was a cornerstone of his life, marked by mutual respect and frequent correspondence during his long absences.

- Married Abigail Smith on October 25, 1764, in Weymouth, Massachusetts.

- Their relationship was characterized by a deep emotional and intellectual connection, evident in over 1,200 surviving letters.
- Lived primarily in Braintree, later moving to Philadelphia and Washington, D.C., during political service.
- Faced long separations due to Adams's diplomatic and political duties, strengthening their bond through written communication.

Children

John and Abigail Adams had five children, four of whom survived to adulthood. Their children's lives reflected the family's commitment to public service, though some faced personal struggles.

- Abigail "Nabby" Adams (1765–1813): Married William Stephens Smith; died of breast cancer.
- John Quincy Adams (1767–1848): Became the sixth U.S. President, served as a diplomat and congressman.
- Susanna Adams (1768–1770): Died in infancy.
- Charles Adams (1770–1800): Struggled with alcoholism, died young of related causes.
- Thomas Boylston Adams (1772–1832): Became a lawyer and judge, and led a quieter life.

Rise to Power

Adams's ascent to prominence was driven by his legal acumen, revolutionary zeal, and diplomatic skill. He emerged as a leading voice in the push for independence and played a critical role in the Continental Congress and abroad.

- Practiced law in Boston, gaining a reputation for defending unpopular causes, including British soldiers in the Boston Massacre trial (1770).

- Elected to the Massachusetts legislature and later to the First Continental Congress (1774).

- Advocated for independence in the Second Continental Congress, helping draft the Declaration of Independence (1776).

- Served as a diplomat in France, the Netherlands, and Britain, securing crucial loans and treaties (1778–1788).

- Elected Vice President under George Washington (1789–1797), then President in 1796.

Influences

Adams was shaped by Enlightenment ideas, Puritan ethics, and his observations of government and human nature. Key figures and experiences molded his political philosophy.

- Inspired by Enlightenment thinkers like John Locke, emphasizing natural rights and government by consent.

- Influenced by his Puritan upbringing, which instilled a sense of duty and moral rigor.

- Mentored by James Otis, whose arguments against British taxation shaped Adams's revolutionary stance.

- Shaped by his legal studies and observations of British governance during diplomatic missions.

Party Affiliation

Adams was a member of the Federalist Party, advocating for a strong central government and economic stability. His affiliation shaped his presidency but also led to conflicts.

- Joined the Federalist Party, aligning with Alexander Hamilton and other proponents of centralized authority.

- Opposed by Democratic-Republicans, led by Thomas Jefferson, who favored states' rights and agrarian interests.

- His Federalist stance created tensions with both allies and opponents, particularly over issues like the Alien and Sedition Acts.

Presidency

Adams served as President from March 4, 1797, to March 4, 1801, navigating a young nation through domestic strife and international tensions. His term was marked by efforts to maintain neutrality and strengthen the government, though it was also fraught with controversy.

- Faced the Quasi-War with France (1798–1800), an undeclared naval conflict sparked by French attacks on American ships.

- Signed the controversial Alien and Sedition Acts (1798), aimed at national security but criticized for curbing free speech.

- Appointed John Marshall as Chief Justice, shaping the judiciary's role in American governance.

- Moved the federal government to Washington, D.C., in 1800, establishing it as the new capital.
- Lost re-election in 1800 to Thomas Jefferson, amid Federalist Party divisions and public discontent.

Accomplishments

John Adams's contributions to the United States were significant, particularly in diplomacy and governance.

- Played a leading role in drafting and advocating for the Declaration of Independence.
- Secured critical Dutch loans during the Revolutionary War, sustaining the American effort.
- Negotiated the Treaty of Paris (1783), ending the Revolutionary War and securing American independence.
- Established the U.S. Navy in 1798 to protect American shipping during the Quasi-War.
- Avoided full-scale war with France through diplomacy, culminating in the Convention of 1800.
- Appointed key judicial figures, including John Marshall, strengthening the Supreme Court's authority.

First Lady's Contributions

Abigail Adams was a trailblazer, influencing policy and advocating for social progress through her partnership with John.

- Advised John on political matters, offering insights on governance, foreign policy, and public sentiment.

- Advocated for women's rights and education, famously urging John to "remember the ladies" in lawmaking.
- Managed the family farm and finances during John's absences, demonstrating resilience and independence.
- Shaped public opinion through her extensive correspondence, promoting republican values.
- Served as a diplomatic hostess in Europe and the U.S., fostering connections with key figures.

Positive Traits

Adams's personal qualities shaped his leadership, often to the nation's benefit.

- Intellectual rigor: His deep knowledge of law and government informed his contributions to the Constitution and foreign policy.
- Principled commitment: His dedication to independence and national sovereignty guided his diplomacy.
- Courage: Defended unpopular causes, like the British soldiers in the Boston Massacre trial, showing integrity.
- Work ethic: Tirelessly served in multiple roles, from congressman to diplomat to president.

Negative Traits

Adams's flaws sometimes hindered his effectiveness and public perception.

- Stubbornness: His rigid adherence to principles alienated allies and fueled political divisions.

- Temper: His prickly demeanor led to conflicts with colleagues, including Alexander Hamilton.
- Sensitivity to criticism: Adams took public and political attacks personally, affecting his decision-making.
- Poor political instincts: His support for the Alien and Sedition Acts damaged his popularity and legacy.

Effects on Presidency

Adams's traits had a mixed impact on his presidency, shaping both successes and setbacks.

- His intellectual clarity and commitment to neutrality prevented war with France but required unpopular decisions.
- Stubbornness and temper strained relations with Congress and his own party, weakening his political support.
- Sensitivity to criticism led to defensive policies like the Alien and Sedition Acts, which tarnished his reputation.
- His principled leadership ensured a stable transition of power to Jefferson, reinforcing democratic norms.

Pets

While less is documented about Adams's pets compared to other presidents, his household included animals typical of their rural lifestyle.

- Kept horses for transportation and farm work, essential for travel and communication.

- Likely had dogs, common in colonial households for protection and companionship.
- Farm animals, such as cows and chickens, supported the family's self-sufficiency in Braintree.

Religious Persuasion

Adams's faith was rooted in New England Puritanism but evolved into a more rational, Unitarian-leaning perspective.

- Raised as a Congregationalist, attending church regularly in his youth.
- Embraced Enlightenment ideas, questioning traditional doctrines while maintaining a belief in divine providence.
- Rejected Trinitarian theology, aligning with Unitarian views that emphasized reason and morality.
- Believed religion should guide personal ethics and public virtue, influencing his governance.

Interesting Anecdotes

Adams's life was filled with colorful moments that reveal his character and the era's challenges.

- Defended British soldiers in the Boston Massacre trial, earning respect for fairness despite public backlash.
- Nicknamed "His Rotundity" as Vice President, a jab at his physique and perceived pompousness.
- Survived a dangerous Atlantic crossing in 1778, braving storms and enemy ships to reach France.

- Exchanged heated letters with Thomas Jefferson, reconciling late in life to restore their friendship.
- Was the first president to live in the White House, moving in November 1800, though it was unfinished.

Ages at Death

John and Abigail Adams lived long lives for their time, passing away in their hometown.

- John Adams died at age 90 on July 4, 1826, in Quincy, Massachusetts.
- Abigail Adams died at age 73 on October 28, 1818, in Quincy, Massachusetts.

Causes of Death

Their deaths reflected the medical limitations of the early 19th century.

- John Adams died of heart failure, likely exacerbated by old age and declining health.
- Abigail Adams died of typhoid fever, contracted during an outbreak in Quincy.

Burial Locations

Both were laid to rest in their hometown, reflecting their deep ties to Massachusetts.

- John Adams: Buried at the United First Parish Church (Church of the Presidents) in Quincy, Massachusetts.
- Abigail Adams: Buried alongside John at the United First Parish Church in Quincy, Massachusetts.

Conclusion

John Adams's legacy as the "Architect of American Independence" lies in his tireless advocacy for liberty, his diplomatic triumphs, and his role in building a stable republic. Despite a presidency marred by controversy and political strife, his contributions to the Declaration of Independence, the Navy, and the judiciary were foundational. Abigail Adams, his equal in intellect and resolve, amplified his impact through her counsel and advocacy. Together, they embodied the revolutionary spirit, leaving an enduring mark on the nation's early history. Their lives, marked by sacrifice and principle, remain a testament to the challenges and triumphs of America's founding era.

Thomas Jefferson: Architect of American Democracy

Introduction

Thomas Jefferson, the third President of the United States, was a towering figure in the founding of the nation, known for his contributions to American ideals of liberty, democracy, and individual rights. As the principal author of the Declaration of Independence, he articulated the principles that defined the American Revolution. His presidency, intellectual pursuits, and complex personal life reflect both his visionary leadership and the contradictions of his era. This summary explores Jefferson's life, presidency, and legacy, highlighting his accomplishments, personal traits, and the contributions of his family.

Early Life

Thomas Jefferson was born on April 13, 1743, at Shadwell plantation in Albemarle County, Virginia. His early years shaped his intellectual curiosity and commitment to public service.

- Grew up in a wealthy planter family, the third of ten children.
- Received a classical education, studying Latin, Greek, and French under private tutors.
- Attended the College of William and Mary from 1760 to 1762, where he developed a passion for law, philosophy, and science.
- Studied law under George Wythe, a prominent Virginia lawyer, fostering his legal and political acumen.
- Inherited land and enslaved people from his father, Peter Jefferson, at age 14, establishing his lifelong connection to agriculture and land ownership.

Family

Jefferson's family life was marked by personal loss and complex relationships, influenced by the social and economic structures of 18th-century Virginia.

- Father, Peter Jefferson, was a surveyor and planter who died when Thomas was 14.
- Mother, Jane Randolph Jefferson, came from a prominent Virginia family but had a distant relationship with her son.
- Married Martha Skelton, a widow, on January 1, 1772, at age 28.

- Maintained a controversial relationship with Sally Hemings, an enslaved woman, after Martha's death, which has been widely debated by historians.
- Owned over 600 enslaved people during his lifetime, including his own children, reflecting the moral contradictions of his era.

Children

Jefferson's children, primarily from his marriage to Martha and his relationship with Sally Hemings, played significant roles in his personal life, though many faced tragedy or societal challenges.

- With Martha Skelton, Jefferson had six children: Martha (1772–1836), Jane (1774–1775), an unnamed son (1777), Mary (1778–1804), Lucy Elizabeth (1780–1781), and Lucy Elizabeth (1782–1784). Only Martha and Mary survived to adulthood.
- With Sally Hemings likely fathered at least six children, four of whom survived to adulthood: Beverly (1798–?), Harriet (1801–?), Madison (1805–1877), and Eston (1808–1856).
- Provided some of his Hemings children with limited freedoms, such as allowing Beverly and Harriet to leave Monticello, though they were not formally emancipated.
- His surviving daughter, Martha, managed Monticello's household after her mother's death and later supported her father's political career.

Rise to Power

Jefferson's ascent to national prominence was driven by his intellectual contributions, legal expertise, and political service during the American Revolution and early republic.

- Began his political career in 1769 as a member of Virginia's House of Burgesses.

- Gained fame as the author of the Declaration of Independence in 1776, at age 33, articulating the principles of liberty and equality.

- Served as Virginia's governor (1779–1781), though his tenure was criticized for ineffective leadership during British invasions.

- Appointed Minister to France (1785–1789), where he developed diplomatic skills and deepened his admiration for Enlightenment ideals.

- Became the first U.S. Secretary of State (1790–1793) under President George Washington, shaping early American foreign policy.

Influences

Jefferson's philosophy and leadership were shaped by Enlightenment thinkers, personal experiences, and the revolutionary spirit of his time.

- Drew inspiration from John Locke, whose ideas on natural rights influenced the Declaration of Independence.

- Admired French philosophers like Montesquieu, whose theories on the separation of powers informed Jefferson's views on government.

- Influenced by his Virginia upbringing, which instilled a belief in agrarianism as the foundation of a virtuous republic.

- His time in France exposed him to revolutionary ideas and refined his vision for American democracy.

- Valued education and intellectual inquiry, founding the University of Virginia as a testament to his belief in knowledge as a public good.

Party Affiliation

Jefferson was a key figure in the development of America's early political party system, advocating for decentralized government and individual liberties.

- Co-founded the Democratic-Republican Party with James Madison, opposing the Federalist Party's centralized policies.

- Believed in states' rights, limited federal power, and an agrarian-based economy, contrasting with Federalist support for commerce and strong national government.

- His party affiliation shaped his rivalry with Alexander Hamilton, influencing debates over the Constitution and economic policy.

- Championed democratic ideals, appealing to farmers, artisans, and small-government advocates, which helped secure his presidential victories.

Presidency

Jefferson served as president from March 4, 1801, to March 4, 1809, during a transformative period in American history. His presidency focused on reducing federal power and expanding the nation's territory.

- Elected in 1800 after a contentious tie with Aaron Burr, resolved by the House of Representatives, marking the first peaceful transfer of power between parties.
- Reduced national debt and government spending, reflecting his belief in frugal governance.
- Faced challenges with the Barbary Wars (1801–1805), authorizing military action to protect American trade.
- Navigated tensions with Britain and France during the Napoleonic Wars, enacting the controversial Embargo Act of 1807 to avoid foreign conflict, which harmed the U.S. economy.
- Strengthened national defense and promoted westward exploration, setting the stage for American expansion.

Accomplishments

Jefferson's achievements as a statesman, writer, and thinker left a lasting impact on the United States.

- Authored the Declaration of Independence, establishing the philosophical foundation for American independence.
- Orchestrated the Louisiana Purchase, doubling the size of the United States and securing its westward expansion.

- Founded the University of Virginia, advancing public education and intellectual freedom.
- Promoted religious freedom through the Virginia Statute for Religious Freedom, a precursor to the First Amendment.
- Reduced federal bureaucracy and taxes, aligning with his vision of limited government.

First Lady's Contributions

Martha Skelton Jefferson died in 1782, before Jefferson's presidency, so there was no official First Lady during his terms. His daughter, Martha Jefferson Randolph, often served as hostess at Monticello and the White House.

- Managed Monticello's household, overseeing domestic affairs and enslaved laborers during Jefferson's frequent absences.
- Hosted social events in Washington, D.C., representing her father with grace and diplomacy.
- Educated her own children at Monticello, preserving Jefferson's intellectual legacy.
- Supported charitable causes, including aid for orphans and the poor in Virginia.
- Maintained family correspondence, strengthening ties with political allies and family members.

Positive Traits and Effects

Jefferson's personal qualities shaped his presidency, often to the benefit of the nation, though not without challenges.

- Intellectual brilliance: His deep knowledge of law, philosophy, and science informed policies like the Louisiana Purchase and educational reforms.

- Visionary leadership: His commitment to democracy and expansion laid the groundwork for America's growth as a nation.

- Diplomatic finesse: His experience in France helped navigate complex foreign relations, avoiding war with European powers.

- Charisma and eloquence: His ability to articulate American ideals inspired public support and unified diverse constituencies.

- These traits enabled bold decisions like territorial expansion but sometimes led to over-idealism, as seen in the economic fallout of the Embargo Act.

Negative Traits and Effects

Jefferson's flaws and contradictions impacted his presidency and legacy, reflecting the complexities of his character.

- Moral inconsistency: His ownership of enslaved people, including his relationship with Sally Hemings, contradicted his advocacy for liberty, tarnishing his moral authority.

- Indecisiveness: His reluctance to confront crises decisively, such as during his governorship or the Embargo Act, led to perceptions of weak leadership.

- Financial mismanagement: Chronic debt from personal spending strained his resources and distracted him from governance.

- Partisan rigidity: His staunch anti-Federalist stance deepened political divisions, complicating national unity.

- These traits contributed to policy failures like the Embargo Act and fueled criticism of his personal and political integrity.

Pets

Jefferson's love for animals, particularly at Monticello, reflected his curiosity about the natural world.

- Kept a mockingbird named Dick, which he taught to sing and perch on his shoulder, providing companionship during writing sessions.

- Owned several dogs, including a sheepdog named Bergere, used to protect livestock at Monticello.

- Received two bear cubs as a gift from explorer Zebulon Pike, which he kept briefly before sending them to a museum, citing their unsuitability as pets.

- Maintained a variety of livestock, including horses and exotic birds, aligning with his interest in agriculture and natural history.

Religious Persuasion

Jefferson's religious beliefs were unconventional for his time, shaped by Enlightenment rationalism and a commitment to religious freedom.

- Identified as a deist, believing in a creator but rejecting organized religion and supernatural doctrines.
- Compiled the Jefferson Bible, editing the New Testament to focus on Jesus' moral teachings, excluding miracles.
- Advocated for the separation of church and state, believing religion was a private matter, as seen in his Virginia Statute for Religious Freedom.
- Respected diverse faiths but was skeptical of institutional clergy, viewing them as potential threats to individual liberty.

Interesting Anecdotes

Jefferson's life was filled with quirky and revealing stories that highlight his personality and era.

- Designed Monticello with innovative features like dumbwaiters and a revolving bookstand, reflecting his love for architecture and invention.
- Kept detailed records of weather and crops, showcasing his scientific curiosity, even during his presidency.
- Smuggled rice seeds from Italy to improve American agriculture, risking legal consequences to benefit farmers.
- Engaged in a playful correspondence with John Adams, reconciling after years of rivalry, with letters discussing everything from politics to aging.

- Once broke his wrist attempting to jump a fence in Paris to impress a married woman, revealing his romantic and impulsive side.

Ages at Death, Causes of Death, and Burial Locations

Jefferson and his wife's lives ended under different circumstances, a reflection of the hardships in their era.

- Thomas Jefferson died on July 4, 1826, at age 83, at Monticello, likely from a combination of infections, kidney failure, and exhaustion. He was buried at Monticello, with an epitaph he wrote highlighting his authorship of the Declaration of Independence and the Virginia Statute for Religious Freedom, and his founding of the University of Virginia.

- Martha Skelton Jefferson died on September 6, 1782, at age 33, at Monticello, likely from complications following childbirth. She was buried at Monticello in the family cemetery, with Jefferson designing her tombstone.

- Their shared burial site at Monticello remains a place of historical significance, visited by thousands annually.

Conclusion

Thomas Jefferson's legacy as the architect of American democracy endures through his authorship of the Declaration of Independence, his vision for an expansive and free nation, and his contributions to education and religious liberty. His presidency, marked by the Louisiana Purchase and a commitment to limited government, shaped the United States' trajectory. Yet, his personal contradictions, particularly regarding slavery, highlight the complexities of his character and the era he lived in. Jefferson's life reflects the aspirations and challenges of a young nation, leaving a profound impact on American history that continues to spark debate and admiration.

James Madison: Father of the Constitution

Introduction

James Madison, the fourth President of the United States, is often called the "Father of the Constitution" for his pivotal role in drafting and promoting the U.S. Constitution and the Bill of Rights. A brilliant political thinker and statesman, Madison's contributions to American governance remain foundational. His presidency, however, faced significant challenges, including the War of 1812, which tested his leadership.

Description

James Madison was a diminutive man, standing at 5 feet 4 inches and weighing about 100 pounds, with a reserved demeanor and sharp intellect. Known for his meticulous preparation and logical arguments, he was less charismatic than his contemporaries but commanded respect through his ideas and writings.

Early Life

Born on March 16, 1751, in Port Conway, Virginia, Madison grew up on his family's plantation, Montpelier. He was the eldest of twelve children in a wealthy, land-owning family. His early education was rigorous, shaped by private tutors and a strong emphasis on classical studies, which prepared him for a life of intellectual and political engagement.

Family

Madison's parents, James Madison Sr. and Nelly Conway Madison, were prominent Virginia gentry. His father was a successful planter and local justice of the peace, while his mother managed the household and supported the family's social standing. Madison's close relationship with his family influenced his sense of duty and responsibility.

Children

James Madison and his wife, Dolley Madison, had no children together. Dolley, however, had a son, John Payne Todd, from her first marriage to John Todd Jr., who died during a yellow fever epidemic in 1793. Madison treated Todd as his own, though Todd's later financial troubles and gambling habits strained the family.

Rise to Power

Madison's political career began in Virginia's revolutionary government, where he served in the state legislature and the Continental Congress. His deep study of political theory and governance led to his central role at the 1787 Constitutional Convention. As a key author of the Federalist

Papers, he advocated for a strong federal government, cementing his national prominence.

Influences

Madison was influenced by Enlightenment thinkers like John Locke and Montesquieu, whose ideas on liberty, separation of powers, and checks and balances shaped his political philosophy. His collaboration with Thomas Jefferson and Alexander Hamilton also molded his views, though he later diverged from Hamilton's Federalist policies.

Party Affiliation

Madison was a founder of the Democratic-Republican Party, alongside Thomas Jefferson. The party opposed the Federalists' centralized government vision, advocating for states' rights, agrarian interests, and limited federal power. Madison's party affiliation defined his political battles and presidency.

Presidency

Madison served as President from 1809 to 1817, during a turbulent period marked by tensions with Britain and France. The War of 1812, often called "Mr. Madison's War," dominated his presidency, with mixed outcomes but a surge in national pride. His administration also saw economic growth and westward expansion.

Accomplishments

- Co-authored the Federalist Papers, crucial for ratifying the Constitution.

- Drafted the Bill of Rights, ensuring individual liberties.
- Led the U.S. through the War of 1812, preserving national sovereignty.
- Supported the establishment of the Second Bank of the United States, stabilizing the economy.
- Promoted infrastructure development, laying the groundwork for national growth.

First Lady's Contributions

- Redefined the role of First Lady by hosting bipartisan social events and fostering political unity.
- Saved key White House artifacts, including a portrait of George Washington, during the British burning of Washington in 1814.
- Promoted American-made goods, boosting national pride and the economy.
- Supported charitable causes, including orphanages, enhancing her public image.
- Maintained extensive correspondence, strengthening political networks for her husband.

Personal Positive Traits

Madison's intellect, diligence, and commitment to republican principles were his greatest strengths. His ability to synthesize complex ideas and negotiate compromises made him a master legislator. These traits helped him navigate constitutional debates and wartime challenges, though his reserved nature sometimes limited his public appeal.

Personal Negative Traits

Madison's lack of charisma and indecisiveness occasionally undermined his leadership. His reluctance to assert executive authority early in the War of 1812 led to military setbacks. His small stature and soft-spoken demeanor also made it harder to inspire public confidence during crises.

Effects on Presidency

Madison's positive traits ensured a strong constitutional legacy and diplomatic successes, like the Treaty of Ghent, which ended the War of 1812. However, his negative traits contributed to initial wartime failures, such as the burning of Washington, and criticism of his leadership style as overly cautious.

Pets

The Madisons kept a green parrot named Polly, which Dolley taught to entertain guests. The bird was a fixture at White House social events and reportedly outlived both James and Dolley, becoming a minor celebrity in Washington.

Religious Persuasion

Madison was raised Episcopalian but adopted a more deistic perspective as an adult, emphasizing reason over dogma. He was a strong advocate for religious freedom, co-authoring Virginia's Statute for Religious Freedom with Jefferson, which influenced the First Amendment.

Interesting Anecdotes

- During the Constitutional Convention, Madison sat near the front, taking detailed notes that became the primary record of the debates, despite his small size making it hard to see over others.

- Dolley Madison served ice cream at White House, a rare treat that popularized the dessert among American elites.

- Madison once fled Washington, D.C., during the British invasion in 1814, earning the nickname "the runaway president," though he quickly returned to rally the nation.

Age at Death

- James Madison died at age 85 on June 28, 1836.
- Dolley Madison died at age 81 on July 12, 1849.

Cause of Death

- James Madison died of heart failure at his Montpelier home, weakened by old age and chronic illness.

- Dolley Madison died of a stroke in Washington, D.C., after years of financial hardship and declining health.

Burial Locations

- James Madison was buried at Montpelier, his family estate in Orange County, Virginia.

- Dolley Madison was buried at Congressional Cemetery in Washington, D.C., but her remains were later reinterred at Montpelier beside her husband in 1858.

Conclusion

James Madison's legacy as the "Father of the Constitution" is unmatched, with his intellectual rigor shaping America's foundational documents. His presidency, while challenged by the War of 1812, reinforced national identity. Dolley Madison's charm and resilience complemented his leadership, leaving a lasting mark on the role of the First Lady. Together, their contributions to the early U.S. government and society reflect a partnership of ideas and action that helped build a nation.

James Monroe: Architect of the Monroe Doctrine

Description

James Monroe, the fifth President of the United States, was a tall, reserved Virginian known for his steady demeanor and commitment to national unity. A Revolutionary War veteran and disciple of Thomas Jefferson, he played a pivotal role in shaping early American foreign policy.

Introduction

James Monroe (1758–1831) served as president from 1817 to 1825, during the "Era of Good Feelings," a period of relative political harmony. Best remembered for the Monroe Doctrine, which warned European powers against further colonization in the Americas, his presidency marked a defining moment in U.S. foreign policy. A lawyer, diplomat, and statesman, Monroe's life reflected the challenges and triumphs of a young nation.

Early Life

Born on April 28, 1758, in Westmoreland County, Virginia, James Monroe grew up in a modest planter family. His early education and experiences shaped his dedication to public service.

- Attended local schools and was tutored by Rev. Archibald Campbell.
- Enrolled at the College of William and Mary in 1774 but left in 1776 to join the Continental Army.
- Fought in key Revolutionary War battles, including Trenton, where he was wounded.
- Studied law under Thomas Jefferson in 1780, forming a lifelong mentorship.

Family

Monroe's family life centered around his wife, Elizabeth Kortright Monroe, and their children. His marriage was a partnership that supported his political career, though Elizabeth's frail health limited her public role.

- Married Elizabeth Kortright in 1786 in New York City.
- Elizabeth was the daughter of a wealthy New York merchant.
- The couple maintained close ties with Virginia and New York elites.
- Lived in various cities, including Philadelphia, Paris, and Washington, D.C., due to Monroe's career.

Children

James and Elizabeth Monroe had three children, though only two survived to adulthood. Their children's lives reflected the family's prominence but also personal tragedies.

- Eliza Monroe Hay (1786–1840): Educated in Paris, married George Hay, and served as White House hostess due to her mother's illness.

- James Spence Monroe (1799–1800): Died in infancy, a loss that deeply affected the family.

- Maria Hester Monroe Gouverneur (1802–1850): Married Samuel Gouverneur in 1820, the first White House wedding.

Rise to Power

Monroe's ascent to the presidency was built on decades of public service, from soldier to diplomat to governor. His loyalty to the Jeffersonian vision earned him broad support.

- Served in the Virginia legislature (1782) and Continental Congress (1783–1786).

- Opposed the Constitution initially but later supported it as a U.S. senator (1790–1794).

- Minister to France (1794–1796), where he navigated tense U.S.-French relations.

- Governor of Virginia (1799–1802, 1811), strengthening the state militia and infrastructure.

- Negotiated the Louisiana Purchase (1803) as a special envoy under Jefferson.
- Secretary of State (1811–1817) and briefly Secretary of War (1814–1815) under Madison.

Influences

Monroe's political philosophy was shaped by mentors, revolutionary ideals, and international experiences.

- Thomas Jefferson instilled republican principles and agrarian values.
- Revolutionary War service fostered a commitment to national independence.
- Time in France exposed him to revolutionary movements and diplomacy.
- George Washington's leadership inspired Monroe's focus on national unity.

Party Affiliation

Monroe was a lifelong Democratic-Republican, advocating for limited federal power and states' rights.

- Aligned with Jefferson and Madison against Federalist policies.
- Supported agrarian interests and westward expansion.
- His presidency saw the decline of the Federalist Party, leading to one-party dominance.

Presidency

Monroe's two terms (1817–1825) were marked by economic challenges, territorial expansion, and the assertion of U.S. sovereignty. The Era of Good Feelings masked underlying sectional tensions.

- Elected in 1816, defeating Federalist Rufus King, and re-elected in 1820 nearly unanimously.
- Faced the Panic of 1819, the first major U.S. economic depression.
- Oversaw the Missouri Compromise (1820), balancing free and slave states.
- Acquired Florida from Spain via the Adams-Onís Treaty (1819).
- Proclaimed the Monroe Doctrine (1823) proclaimed, opposing European intervention in the Americas.
- Supported internal improvements, like roads and canals, despite constitutional concerns.

Accomplishments

- Negotiated the Louisiana Purchase, doubling U.S. territory.
- Established the Monroe Doctrine, a cornerstone of U.S. foreign policy.
- Secured Florida and defined U.S.-Spanish boundaries through the Adams-Onís Treaty.
- Brokered the Missouri Compromise, delaying sectional conflict.

- Promoted national infrastructure, laying the groundwork for economic growth.

First Lady's Contributions

- Hosted diplomatic receptions, strengthening U.S. ties with foreign dignitaries.
- Renovated the White House after its 1814 burning, enhancing its prestige.
- Supported education for women, influencing her daughter, Eliza's schooling in Paris.

Positive Traits

Monroe's strengths shaped a steady, unifying presidency.

- Diplomacy: His experience abroad fostered effective foreign policy.
- Integrity: Known for honesty, he earned trust across political factions.
- Dedication: Tireless commitment to public service strengthened national institutions.
- These traits enabled landmark achievements like the Monroe Doctrine and territorial expansion.

Negative Traits

Monroe's weaknesses occasionally hindered his leadership.

- Indecisiveness: Slow to act during the Panic of 1819, prolonging economic woes.

- Passivity: Relied heavily on advisors, limiting bold domestic reforms.
- These traits contributed to unresolved sectional tensions and economic challenges.

Pets

The Monroe family kept pets, reflecting their domestic life.

- Owned a spaniel named Buddy, a gift from a Virginia friend.
- Elizabeth kept canaries in the White House, enjoying their songs.
- Pets provided comfort during frequent relocations.

Religious Persuasion

Monroe's faith was private but influenced his moral outlook.

- Raised in the Anglican (later Episcopal) Church, common among Virginia elites.
- Attended services irregularly due to public duties and Elizabeth's health.
- Expressed deist-leaning views, emphasizing reason and morality over dogma.

Interesting Anecdotes

- Nearly fought a duel with Alexander Hamilton in 1797 over a political dispute, averted by diplomacy.
- Wore Revolutionary War-era clothing during his 1817 goodwill tour, earning the nickname "Last Cocked Hat."

- Hosted Lafayette in 1824, a tearful reunion of Revolutionary War allies.
- Lost his Virginia estate due to debts, living modestly in New York after his presidency.

Age at Death

- James Monroe died at 73 on July 4, 1831.
- Elizabeth Monroe died at 62 on September 23, 1830.

Cause of Death

- James Monroe succumbed to heart failure and possibly tuberculosis.
- Elizabeth Monroe died of an unspecified illness, likely related to chronic health issues.

Burial Location

- James Monroe was buried in Marble Cemetery, New York City; reinterred in 1858 at Hollywood Cemetery, Richmond, Virginia.
- Elizabeth Monroe was buried in Marble Cemetery, New York City, and reinterred with James in Richmond, Virginia.

Conclusion

James Monroe's life embodied the spirit of a fledgling nation striving for identity and influence. His Monroe Doctrine defined U.S. foreign policy for centuries, while his steady leadership bridged partisan divides. Despite economic and sectional challenges, his accomplishments—territorial expansion, diplomatic triumphs, and national unity—

cemented his legacy. Elizabeth's quiet grace supported his public life, though her contributions were curtailed by illness. Monroe's story, from Revolutionary soldier to elder statesman, reflects the resilience and ambition of early America.

John Quincy Adams: Architect of American Diplomacy

Introduction

John Quincy Adams, the sixth President of the United States, was a towering figure in early American history, renowned for his contributions to diplomacy and his unwavering commitment to public service. Born into a family steeped in revolutionary ideals, Adams carved a legacy as a statesman, diplomat, and president whose influence shaped the nation's foreign policy and domestic principles. His presidency, though marked by challenges, reflected his dedication to national unity and progress, cementing his reputation as a principled leader and the architect of American diplomacy.

Early Life

John Quincy Adams was born on July 11, 1767, in Braintree (now Quincy), Massachusetts, to John Adams, a future president, and Abigail Adams, a formidable intellectual. His early years were shaped by the American Revolution, exposing him to the ideals of liberty and governance.

- Grew up in a household immersed in political discourse, witnessing key events of the Revolution.

- Accompanied his father on diplomatic missions to Europe at age 10, gaining early exposure to international affairs.

- Educated informally in Europe, studying in France and the Netherlands, and later attended Harvard College, graduating in 1787.

- Studied law and was admitted to the bar in 1790, beginning a career that blended law and public service.

Family

Adams' family was a cornerstone of his life, with his parents instilling values of duty and intellect. His marriage to Louisa Catherine Johnson further enriched his personal and public life.

- Son of John Adams, second U.S. President, and Abigail Adams, a noted letter-writer and advocate for women's rights.

- Married Louisa Catherine Johnson in 1797, a cultured and resilient woman born in London to an American father and British mother.

- Maintained close ties with siblings, including his brother Charles, despite personal tragedies in the family.

Children

John Quincy and Louisa Adams had four children, though their family life was marked by loss and separation due to Adams' diplomatic duties.

- George Washington Adams (1801–1829): Struggled with personal issues and died young, possibly by suicide.

- John Adams II (1803–1834): Worked in business but faced health and financial difficulties, dying early.

- Charles Francis Adams (1807–1886): Became a successful diplomat and congressman, carrying on the family's political legacy.

- Louisa Catherine Adams (1811–1812): Died in infancy during the family's time in Russia.

Rise to Power

Adams' ascent to the presidency was built on a remarkable career in diplomacy and politics, showcasing his intellect and dedication to public service.

- Began diplomatic career at age 14 as a secretary to the U.S. minister to Russia.

- Served as minister to the Netherlands (1794–1797), Prussia (1797–1801), Russia (1809–1814), and Britain (1815–1817).

- Played a key role in negotiating the Treaty of Ghent (1814), ending the War of 1812.

- Served as Secretary of State under President James Monroe (1817–1825), authoring the Monroe Doctrine.

- Elected president in 1824 through a controversial "corrupt bargain" in the House of Representatives after no candidate won a majority in the Electoral College.

Influences

Adams' worldview was shaped by his family, education, and extensive diplomatic experience, which emphasized national interest and moral governance.

- Inspired by his father's revolutionary ideals and his mother's advocacy for education and ethics.
- Influenced by Enlightenment thinkers like John Locke, emphasized reason and republicanism.
- Shaped by European diplomatic experiences, fostering a belief in strong, principled foreign policy.
- Guided by a sense of duty to promote American unity and economic growth.

Party Affiliation

Adams' political affiliations evolved, reflecting the fluid party dynamics of his era and his commitment to principle over partisanship.

- Began as a Federalist, like his father, supporting a strong national government.
- Shifted to the Democratic-Republican Party by the early 1800s, aligning with Jeffersonian ideals.
- As president, aligned with the National Republican Party, advocating for federal infrastructure and economic development.

- Later, as a congressman, associated with the Whig Party, focusing on anti-slavery and national unity.

Presidency

Adams served as president from March 4, 1825, to March 4, 1829, in a single term marked by ambitious goals but political opposition.

- Faced a hostile Congress dominated by Andrew Jackson's supporters, limiting his legislative success.
- Proposed a bold agenda for national infrastructure, including roads, canals, and a national university.
- Advocated for protective tariffs to boost American industry, laying the groundwork for economic growth.
- Navigated foreign policy challenges, strengthening U.S. relations with Latin America and Europe.
- Lost re-election in 1828 to Andrew Jackson in a bitterly contested campaign.

Accomplishments

John Quincy Adams' achievements spanned diplomacy, governance, and post-presidential advocacy, leaving a lasting impact on American policy.

- Authored the Monroe Doctrine (1823), establishing U.S. opposition to European colonialism in the Americas.
- Negotiated key treaties, including the Treaty of Ghent, securing peace with Britain.
- Promoted internal improvements like the Erie Canal, fostering economic development.

- As a congressman after his presidency, fiercely opposed slavery, earning the nickname "Old Man Eloquent."
- Defended the Amistad captives (1841), securing their freedom in a landmark Supreme Court case.

First Lady's Contributions

Louisa Catherine Adams, the first foreign-born First Lady, brought grace and resilience to her role despite personal hardships.

- Hosted social events to foster political alliances, enhancing the administration's diplomatic efforts.
- Advocated for women's education and social issues, reflecting her intellectual upbringing.
- Wrote memoirs that provided valuable insights into early American political life.
- Supported her husband's career through personal sacrifices, including long separations during his diplomatic postings.

Positive Traits

Adams' strengths shaped his principled leadership but sometimes hindered his political effectiveness.

- Intellectual brilliance: A master of diplomacy and policy, with a deep understanding of governance.
- Unwavering integrity: Refused to compromise principles for political gain, earning respect as a statesman.

- Dedication to public service: Served in multiple roles, from diplomat to congressman, over decades.
- Linguistic and cultural fluency: Spoke multiple languages, aiding his diplomatic success.

Negative Traits

Adams' flaws contributed to challenges during his presidency, particularly in navigating political opposition.

- Aloof demeanor: Struggled to connect with the public, appearing distant and elitist.
- Political inflexibility: Refused to engage in patronage, alienating potential allies.
- Poor political instincts: Misjudged the growing populist sentiment, leading to his 1828 defeat.
- Tendency toward overwork: His relentless work ethic sometimes strained family and personal health.

Effects on Presidency

Adams' traits profoundly influenced his presidency, creating both opportunities and obstacles.

- His intellectual vision led to ambitious proposals for national development, but his aloofness hindered public support.
- Integrity prevented corruption but limited his ability to build coalitions in a fractious Congress.
- His diplomatic expertise strengthened U.S. foreign policy, but domestic political missteps undermined his agenda.

- His inflexibility exacerbated tensions with Jacksonians, leading to legislative gridlock.

Pets

While less documented than other aspects of his life, Adams' household included pets that reflected the era's simplicity.

- Kept a pet alligator during his presidency, reportedly housed in a White House bathtub, a gift from the Marquis de Lafayette.
- Likely had dogs and other common pets, though specific records are sparse.

Religious Persuasion

Adams' faith was a private but guiding force in his life, rooted in New England's Puritan traditions.

- Raised in the Congregationalist Church, reflecting his family's Puritan heritage.
- Later aligned with Unitarianism, embracing a rational, less dogmatic approach to Christianity.
- Believed in moral governance and divine providence, influencing his anti-slavery stance.
- Attended church regularly but kept his faith private, focusing on ethical conduct over public displays.

Interesting Anecdotes

Adams' life was filled with unique stories that highlight his character and era.

- As a boy, watched the Battle of Bunker Hill from a hilltop near his home, shaping his patriotic fervor.
- Kept a diary from age 12 until his death, one of the most detailed personal records of any American president.
- Swam daily in the Potomac River during his presidency, once nearly drowning when his clothes became waterlogged.
- Defended free speech in Congress by presenting anti-slavery petitions, defying gag rules despite threats.

Ages at Death

John Quincy and Louisa Adams lived long lives marked by service and personal loss.

- John Quincy Adams died at age 80 on February 23, 1848.
- Louisa Catherine Adams died at age 77 on May 15, 1852.

Causes of Death

Both faced health challenges in their later years, typical of the 19th century.

- John Quincy Adams suffered a stroke on the House floor and died two days later in the Capitol.
- Louisa Catherine Adams died of a heart attack, weakened by years of poor health and personal grief.

Burial Locations

The couple's final resting places reflect their ties to Massachusetts and their family legacy.

- John Quincy Adams is buried at the United First Parish Church (Church of the Presidents) in Quincy, Massachusetts, alongside his parents.

- Louisa Catherine Adams is also buried at the United First Parish Church in Quincy, Massachusetts.

Conclusion

John Quincy Adams' legacy as the architect of American diplomacy endures through his contributions to foreign policy, national development, and the fight against slavery. His presidency, though constrained by political opposition, reflected his vision for a stronger, united America. Louisa Adams complemented her husband's efforts with her social acumen and resilience, navigating the challenges of a public life. Together, they embodied a commitment to duty and principle, leaving an indelible mark on the nation's early history. Adams' post-presidential career in Congress, particularly his anti-slavery advocacy, further solidified his reputation as a statesman of enduring moral courage.

Andrew Jackson: The Hero of New Orleans

Introduction

Andrew Jackson, the seventh President of the United States, served from 1829 to 1837 and remains one of the most polarizing figures in American history. Known as the "Hero of New Orleans" for his decisive victory in the Battle of New Orleans during the War of 1812, Jackson was a champion of the common man, a fierce advocate for American expansion, and a controversial leader whose policies shaped the nation's trajectory. His presidency marked the rise of Jacksonian democracy, emphasizing populist ideals and strong executive power, while also leaving a legacy of contentious decisions, particularly regarding Native American removal and economic policy.

Early Life

Andrew Jackson was born on March 15, 1767, in the Waxhaws region on the border of North and South Carolina, to Scotch-Irish immigrants. His early life was marked by hardship and conflict, shaping his resilient and combative nature.

- Grew up in a frontier settlement with limited formal education, though he later studied law.
- Orphaned by age 14 after his father died before his birth, and his mother and brothers died during the Revolutionary War.
- Served as a courier in the Revolutionary War at age 13, captured by British forces, and endured harsh treatment, fostering a lifelong hatred of the British.
- Developed a reputation for toughness, surviving smallpox and a sword slash from a British officer for refusing to clean boots.

Family

Jackson's family life centered around his marriage to Rachel Donelson Robards, whom he wed in 1794 after a tumultuous courtship. His immediate family was small, defined by tragedy and loyalty.

- Mother, Elizabeth Hutchinson Jackson, instilled in him a sense of independence and moral duty before her death in 1781.
- Had two older brothers, Hugh and Robert, both of whom died during the Revolutionary War.
- Married Rachel, who faced public scandal due to an earlier marriage not fully dissolved at the time of their union, a controversy that haunted Jackson.

- No surviving biological siblings or parents by adulthood, making his wife and adopted kin central to his personal life.

Children

Andrew and Rachel Jackson had no biological children but built a family through adoption and guardianship.

- Adopted Andrew Jackson Jr., the son of Rachel's brother, Severn Donelson, in 1809, raising him as their own.

- Became legal guardian to several children, including Lyncoya, a Creek Indian orphan found during the Creek War in 1813, whom Jackson raised at the Hermitage.

- Also cared for the children of relatives and friends, such as Andrew Jackson Donelson, who later served as Jackson's private secretary during his presidency.

- Jackson's devotion to his adopted family reflected his desire for a legacy, though he faced challenges balancing public life with family responsibilities.

Rise to Power

Jackson's ascent to prominence was driven by military success, legal practice, and political ambition in the frontier state of Tennessee.

- Became a lawyer in 1787, building a reputation in Nashville as a prosecutor and land speculator.

- Served as Tennessee's first U.S. Representative (1796) and briefly as a U.S. Senator (1797–1798) before resigning.

- Gained national fame as a military leader in the War of 1812, particularly for his victory at the Battle of New Orleans in 1815, earning the nickname "Old Hickory."

- Led campaigns against Native American tribes, including the Creek War (1813–1814) and First Seminole War (1817–1818), enhancing his reputation as a decisive leader.

- Appointed territorial governor of Florida in 1821, solidifying his influence in the expanding United States.

Influences

Jackson's worldview was shaped by his frontier upbringing, wartime experiences, and democratic ideals.

- The Revolutionary War instilled a deep patriotism and distrust of centralized authority, particularly British-style aristocracy.

- Frontier life fostered a belief in self-reliance and the rights of the common man over elites.

- Mentors like John Overton, a Tennessee lawyer, guided his early legal and political career.

- His wife, Rachel, provided emotional support and influenced his personal values, though her death in 1828 deepened his bitterness toward political enemies.

- Admired the democratic principles of Thomas Jefferson, adapting them to a more populist vision.

Party Affiliation

Jackson was a key figure in the founding of the Democratic Party, which emerged from the Democratic-Republican Party during his era.

- Initially aligned with the Democratic-Republican Party, opposing Federalist policies of centralized power.
- By the 1828 election, his supporters formed the Democratic Party, emphasizing states' rights, limited government, and populist appeal.
- Opposed the Whig Party, led by rivals like Henry Clay, which favored a strong national bank and federal infrastructure projects.
- His party affiliation solidified a political divide, with Democrats championing the interests of farmers, laborers, and frontiersmen.

Presidency

Jackson's presidency (1829–1837) was transformative, marked by bold executive actions and divisive policies that reshaped American governance.

- Strengthened the presidency by using the veto power extensively, most notably against the Second Bank of the United States in 1832.
- Faced the Nullification Crisis (1832–1833), asserting federal authority against South Carolina's attempt to nullify tariffs, balancing states' rights with national unity.

- Implemented the Indian Removal Act of 1830, leading to the Trail of Tears, forcibly relocating Native American tribes, a policy now widely condemned.

- Paid off the national debt in 1835, a rare achievement, but his Specie Circular and bank policies contributed to the Panic of 1837.

- Expanded voting rights for white males by reducing property requirements, advancing democratic participation.

Accomplishments

- Defeated British forces at the Battle of New Orleans (1815), boosting American morale and national pride.

- Eliminated the national debt in 1835, the only time in U.S. history.

- Vetoed the recharter of the Second Bank of the United States, curbing its influence and appealing to anti-elite sentiments.

- Strengthened the executive branch through assertive use of presidential powers, setting a precedent for future leaders.

- Expanded democratic participation by advocating for broader suffrage among white males.

First Lady's Contributions

- Rachel Jackson died in December 1828, before Andrew's inauguration, and thus never served as First Lady.

- Emily Donelson, Jackson's niece, acted as White House hostess, managing social events with grace despite political tensions.
- Sarah Yorke Jackson, Andrew Jackson Jr.'s wife, later served as co-hostess, maintaining the Hermitage and supporting family interests.
- Rachel's memory influenced Jackson's presidency, as he dedicated efforts to defend her honor against slander from political opponents.

Positive Traits and Effects

Jackson's personal qualities profoundly shaped his leadership style and presidency.

- Courage and decisiveness, evident in his military victories, enabled bold policy moves like the bank veto.
- Charisma and connection with common people strengthened his political base, fostering Jacksonian democracy.
- Loyalty to allies and family inspired fierce devotion, building a strong network of supporters.
- Determination allowed him to overcome personal and political obstacles, such as surviving assassination attempts.
- These traits enhanced his ability to push through reforms and rally public support, though they sometimes led to inflexibility.

Negative Traits and Effects

Jackson's flaws contributed to some of his presidency's most criticized actions.

- Stubbornness and vindictiveness fueled conflicts, such as his feud with the Second Bank, destabilizing the economy.

- Prejudice against Native Americans drove the Indian Removal Act, causing immense suffering.

- Impulsiveness led to decisions like the Specie Circular, which exacerbated economic crises.

- Tendency to personalize political disputes, such as the Eaton Affair, disrupted his administration and alienated allies.

- These traits often deepened divisions, undermining national unity and leaving a mixed legacy.

Pets

Jackson was not widely known for keeping pets, but animals played a role in his life at the Hermitage, his Tennessee plantation.

- Owned horses, including his prized warhorse, Sam Patch, used during military campaigns.

- Likely kept dogs and farm animals typical of a 19th-century plantation, though specific records are scarce.

- His household included animals for practical purposes, reflecting his rural lifestyle rather than sentimental attachment.

Religious Persuasion

Jackson's faith evolved over time, influenced by his wife and personal experiences.

- Raised in a Presbyterian household but not deeply religious in youth, focusing more on frontier survival.
- Rachel's devout Presbyterianism encouraged his later embrace of Christianity.
- Joined the Presbyterian Church after Rachel's death in 1828, finding solace in faith during his presidency.
- His religious beliefs informed his moral rhetoric but did not heavily shape policy, except in justifying expansionist ideals like Manifest Destiny.

Interesting Anecdotes

Jackson's life was filled with colorful and dramatic moments that reflect his fiery personality.

- Survived the first known assassination attempt on a U.S. president in 1835 when an assailant's pistols misfired; Jackson chased the attacker with his cane.
- Fought in numerous duels, including one in 1806 where he killed Charles Dickinson, suffering a bullet wound that remained in his chest for life.
- Adopted Lyncoya, a Creek orphan, during the Creek War, an unusual act for the time, showing a complex attitude toward Native Americans.

- Hosted raucous inaugural celebrations in 1829, where crowds mobbed the White House, forcing Jackson to escape through a window.

- Kept a parrot named Poll, which reportedly swore at Rachel's funeral, adding a humorous footnote to his legacy.

Ages at Death, Causes of Death, and Burial Locations

- **Andrew Jackson**: Died at age 78 on June 8, 1845, at the Hermitage in Nashville, Tennessee. The cause of death was chronic tuberculosis, dropsy (edema), and heart failure, likely exacerbated by old wounds and poor health. Buried at the Hermitage alongside Rachel in the garden.

- **Rachel Jackson**: Died at age 61 on December 22, 1828, in Nashville, Tennessee, before Jackson's inauguration. The cause of death was a heart attack, possibly linked to stress from public scandals about her marriage. Buried at the Hermitage in the garden, where Jackson later joined her.

Conclusion

Andrew Jackson's life and presidency embody the contradictions of early America: a champion of democratic ideals for white men, yet a figure whose policies inflicted profound harm, particularly on Native Americans. His military triumphs, like the Battle of New Orleans, cemented his heroic status, while his presidency reshaped the nation through populist reforms and assertive leadership. However, his role in the Trail of Tears and economic turmoil remains a dark stain on his legacy. Jackson's personal grit, loyalty, and flaws defined an era of expansion and division, leaving a complex legacy as the "Hero of New Orleans" who both united and fractured a growing nation.

Martin Van Buren: Architect of the Democratic Party

Description

Martin Van Buren, the eighth President of the United States, was a skilled politician known for his strategic mind and ability to build political coalitions. Often called the "Little Magician" for his political finesse, he played a pivotal role in shaping the modern Democratic Party. His presidency, however, was marred by economic challenges that defined much of his legacy.

Introduction

Martin Van Buren served as president from 1837 to 1841, navigating a turbulent period in American history. Born into modest circumstances, he rose through New York politics to become a national figure. His career was marked by his organizational genius and commitment to limited government, but his presidency faced significant economic crises that overshadowed his achievements.

Early Life

Martin Van Buren was born on December 5, 1782, in Kinderhook, New York, a small Dutch-American community. His early life shaped his pragmatic and resourceful character.

- Grew up in a tavern owned by his father, Abraham Van Buren, exposing him to diverse social interactions.
- Attended local schools, including Kinderhook Academy, but had limited formal education.
- Began studying law at age 14 under attorney Francis Sylvester, developing a keen interest in politics.

Family

Van Buren's family ties were rooted in Dutch-American culture, influencing his early worldview.

- Father, Abraham, was a farmer, tavern keeper, and Revolutionary War veteran.
- Mother, Maria Hoes Van Alen, was a widow with three children before marrying Abraham.
- Had five siblings and three half-siblings from his mother's prior marriage.
- Married Hannah Hoes, his childhood sweetheart and distant cousin, on February 21, 1807.

Children

Martin and Hannah Van Buren had five children, four of whom survived to adulthood.

- Abraham Van Buren (1807–1873): Served as his father's private secretary during the presidency.

- John Van Buren (1810–1866): Became a prominent New York lawyer and politician.

- Martin Van Buren Jr. (1812–1855): Assisted his father with political correspondence.

- Winfield Scott Van Buren (1814–1814): Died in infancy.

- Smith Thompson Van Buren (1817–1876): Edited his father's papers later in life.

Rise to Power

Van Buren's ascent was driven by his political acumen and alliances in New York's competitive political landscape.

- Admitted to the bar in 1803, practicing law in Kinderhook.

- Elected to the New York State Senate in 1812, aligning with the Democratic-Republican Party.

- Became New York's Attorney General in 1815, strengthening his influence.

- Formed the Albany Regency, a political machine that controlled New York politics.

- Served as U.S. Senator (1821–1828), Governor of New York (1829), and Secretary of State under Andrew Jackson (1829–1831).

- Appointed Vice President under Jackson (1833–1837), positioning him for the presidency.

Influences

Van Buren's political philosophy was shaped by key figures and ideologies.

- Admired Thomas Jefferson's principles of limited government and states' rights.
- Influenced by Andrew Jackson's populism, Van Buren preferred a structured party organization.
- Drew from Dutch-American values of thrift and community, emphasizing practical governance.

Party Affiliation

Van Buren was a founder of the Democratic Party, formalizing its structure in the 1820s.

- Initially a Democratic-Republican, he broke with factions to form the Democratic Party with Jackson.
- Championed party discipline through the Albany Regency, setting a model for national politics.
- Advocated for a strong party system to counter elitist governance, shaping modern political campaigns.

Presidency

Van Buren's presidency (1837–1841) was dominated by the Panic of 1837, an economic depression that defined his term.

- Took office amid financial instability caused by Jackson's banking policies.

- Established the Independent Treasury System to separate government funds from private banks.
- Faced criticism for his handling of the economy and Indian removal policies, including the Trail of Tears.
- Lost re-election in 1840 to William Henry Harrison, reflecting public discontent.

Accomplishments

- Established the Independent Treasury System, stabilizing federal finances.
- Strengthened the Democratic Party's organizational structure, influencing future politics.
- Maintained peace with Britain during border disputes, avoiding war.
- Promoted fiscal restraint, aligning with his limited government philosophy.

First Lady's Contributions

Hannah Hoes Van Buren died in 1819, before Van Buren's presidency, so there was no First Lady during his term. His daughter-in-law, Angelica Singleton Van Buren, served as White House hostess.

- Angelica organized formal social events, enhancing diplomatic relations.
- Introduced European-style elegance to White House functions, elevating its prestige.
- Supported Van Buren's political image through gracious hospitality.

Positive Traits

Van Buren's strengths shaped his political success but had mixed effects on his presidency.

- Strategic and diplomatic, he built coalitions effectively.
- Disciplined and organized, he created enduring political structures.
- Charismatic and persuasive, he won allies but struggled to inspire the public during crises.

Negative Traits

Van Buren's weaknesses hindered his ability to address national challenges.

- Perceived as aloof, distancing him from voters during economic hardship.
- Overly cautious, delaying decisive action on the Panic of 1837.
- Reliance on party loyalty limited his appeal to a broader electorate, contributing to his 1840 defeat.

Pets

Little is documented about Van Buren's pets, reflecting his focus on political rather than personal details.

- No specific records exist of pets during his presidency or personal life.
- Typical of the era, any household animals would likely have been utilitarian, such as horses or farm animals.

Religious Persuasion

Van Buren was raised in the Dutch Reformed Church, reflecting his cultural roots.

- Maintained a private faith, rarely discussing religion publicly.
- Attended Dutch Reformed services but was not overtly religious in political life.
- His faith emphasized personal morality and community, aligning with his governance style.

Interesting Anecdotes

Van Buren's life included notable moments that highlight his personality and era.

- Nicknamed "Old Kinderhook," which inspired the term "OK" in American slang.
- Wore elegant clothing, earning the moniker "Dandy" from critics who viewed him as elitist.
- Hosted lavish White House dinners, contrasting with his frugal personal habits.

Age at Death

- Martin Van Buren died at age 79 on July 24, 1862.
- Hannah Hoes Van Buren died at age 35 on February 5, 1819.

Cause of Death

- Martin Van Buren died of bronchial asthma and heart failure at his Lindenwald estate in Kinderhook.

- Hannah Hoes Van Buren died of tuberculosis, a common illness at the time.

Burial Location

- Martin Van Buren is buried at Kinderhook Reformed Church Cemetery, Kinderhook, New York.
- Hannah Hoes Van Buren is buried at the same cemetery, alongside her husband.

Conclusion

Martin Van Buren's legacy as the architect of the Democratic Party endures, despite a presidency overshadowed by economic turmoil. His political innovations laid the groundwork for modern party politics, but his cautious leadership struggled against the Panic of 1837. A man of contrasts—strategic yet aloof, principled yet pragmatic—Van Buren's life reflects the complexities of early American governance. His contributions, alongside Angelica's social grace, shaped a pivotal era, leaving a lasting imprint on the nation's political landscape.

William Henry Harrison: The Briefest Presidency

Introduction

William Henry Harrison, the ninth President of the United States, holds a unique place in American history due to his extraordinarily short presidency, lasting only 31 days. Known for his military heroism and political leadership, Harrison's life was marked by service to his country, from his early days as a soldier to his time as a territorial governor and senator. His presidency, however, was cut short by illness, leaving a legacy defined more by potential than accomplishment. This summary explores Harrison's life, family, political career, and the fleeting impact of his presidency, alongside notable aspects of his personal life and the contributions of his wife, Anna Harrison.

Description

William Henry Harrison was a military officer, politician, and statesman whose career spanned the early years of the United States. Born into a prominent Virginia family, he rose to national prominence through his military victories, particularly at the Battle of Tippecanoe, earning the nickname "Old Tippecanoe." His brief presidency in 1841 was a milestone as the first Whig administration, but his death shortly after taking office left his agenda unfulfilled.

Early Life

Born on February 9, 1773, at Berkeley Plantation in Charles City County, Virginia, Harrison was the youngest of seven children in a wealthy, politically connected family. His father, Benjamin Harrison V, was a signer of the Declaration of Independence and a governor of Virginia. William grew up in a privileged environment, educated by private tutors. He briefly attended Hampden-Sydney College before studying medicine at the University of Pennsylvania, though he abandoned this path to pursue a military career, joining the U.S. Army in 1791 at age 18.

Family

Harrison married Anna Tuthill Symmes in 1795, a union that lasted until his death. Anna, born in 1775 in New Jersey, was the daughter of a wealthy landowner and judge. The couple faced initial opposition from Anna's father, who questioned Harrison's prospects, but their marriage proved strong and enduring. They settled in North Bend, Ohio, where they built a family and a life centered on public service.

Children

William and Anna Harrison had ten children, six sons and four daughters, though not all survived to adulthood:

- Elizabeth Bassett (1796–1846)

- John Cleves Symmes (1798–1830)
- Lucy Singleton (1800–1826)
- William Henry Jr. (1802–1838)
- John Scott (1804–1878), father of future President Benjamin Harrison
- Benjamin (1806–1840)
- Mary Symmes (1809–1842)
- Carter Bassett (1811–1839)
- Anna Tuthill (1813–1865)
- James Findlay (1814–1817)
  Their large family faced tragedy, with several children predeceasing their parents, and their surviving descendants continued the Harrison legacy, notably through grandson Benjamin Harrison, the 23rd U.S. President.

Rise to Power

Harrison's path to prominence began with his military service. Commissioned as an ensign, he served under General Anthony Wayne in the Northwest Indian War, distinguishing himself at the Battle of Fallen Timbers in 1794. In 1798, he left the army to enter politics, becoming secretary of the Northwest Territory and later its delegate to Congress. Appointed governor of the Indiana Territory in 1801, he served for 12 years, negotiating treaties with Native American tribes and leading forces to victory at the Battle of Tippecanoe in 1811. This military success propelled him to national fame. After serving as a general in the War of 1812, he entered national politics as a U.S. Representative, Ohio state senator, and U.S. Senator. His 1840 presidential campaign, built on his "Tippecanoe and Tyler Too" slogan, capitalized on his military hero image, leading to a landslide victory over incumbent Martin Van Buren.

Influences

Harrison's worldview was shaped by his Virginia upbringing, his father's revolutionary ideals, and his frontier experiences. His military mentors, like General Wayne, instilled discipline and strategic thinking. As territorial governor, he was influenced by Federalist principles of strong governance but adapted to the populist sentiments of the emerging West. His negotiations with Native American tribes reflected a pragmatic, often controversial approach to expansion, balancing diplomacy with military force. His Whig Party affiliation aligned him with advocates of economic development and limited executive power, though his brief presidency limited his ability to enact these ideals.

Party Affiliation

Harrison was a member of the Whig Party, a coalition formed in the 1830s to oppose Andrew Jackson's Democratic policies. The Whigs favored a strong Congress, economic modernization, and infrastructure development. Harrison's campaign avoided detailed policy discussions, focusing instead on his military record and populist appeal, which unified diverse Whig factions. His affiliation with the Whigs positioned him as a counterpoint to Democratic populism, though his short tenure prevented significant party-driven governance.

Presidency

Inaugurated on March 4, 1841, Harrison's presidency lasted only 31 days, ending with his death on April 4, 1841. His administration had little time to enact policies, but he aimed to reform the spoils system, strengthen the economy through a national bank, and limit executive power. His inaugural address, the longest in history at over 8,000 words, outlined a Whig agenda but was delivered in cold, rainy weather, possibly contributing to his fatal illness. Harrison's death led to a constitutional crisis over succession, ultimately resolved when Vice President John Tyler assumed the presidency.

Accomplishments

Due to his brief tenure, Harrison's presidential accomplishments were limited:

- Delivered a comprehensive inaugural address outlining Whig priorities, emphasizing congressional authority and economic reform.

- Appointed key cabinet members, including Daniel Webster as Secretary of State, to advance Whig policies.

- Issued calls for a special session of Congress to address economic issues, though he did not live to oversee it.

Contributions of the First Lady

Anna Harrison never served as First Lady in Washington, D.C., due to her husband's death. She was preparing to join him when he fell ill. Her contributions were primarily domestic and familial:

- Raised ten children, managing the family's Ohio estate during Harrison's frequent absences.
- Supported her husband's career through hospitality and maintaining political connections in Ohio.
- Advocated for education and community welfare, influencing her children's public service, notably her son John Scott and grandson Benjamin Harrison.

Positive and Negative Personality Traits

Harrison's personality shaped his brief presidency and public image:

- **Positive Traits**: Charismatic, courageous, and dedicated, Harrison's military leadership and approachable demeanor won him widespread admiration. His commitment to public service and ability to connect with common voters fueled his electoral success.
- **Negative Traits**: His tendency to micromanage and his verbosity, evident in his lengthy inaugural address, may have strained his health and limited his effectiveness. His reluctance to delegate during his campaign and early presidency led to overwork, possibly exacerbating his illness.

These traits had little time to impact his presidency directly, but his charisma solidified his campaign's success, while his overexertion may have contributed to his demise.

Pets

Harrison was not known to have kept pets during his presidency, likely due to its brevity. However, as a farmer and frontier governor, he likely had working animals such as horses and livestock at his North Bend estate. Specific records of personal pets are absent, reflecting the limited documentation of his brief time in office.

Religious Persuasion

Harrison was raised in the Anglican (Episcopal) Church, common among Virginia's elite. Later in life, he and Anna attended Presbyterian services in Ohio, reflecting the influence of frontier religious communities. His faith was personal but not overtly public, and he did not emphasize religion in his political career, aligning with the era's separation of church and state.

Interesting Anecdotes

- During his 1840 campaign, Harrison's team crafted a populist image, portraying him as a log cabin-dwelling, cider-drinking everyman, despite his aristocratic background. This "Log Cabin and Hard Cider" campaign was a masterful use of propaganda.

- At the Battle of Tippecanoe, Harrison reportedly rode a white horse, making him a visible target, yet emerged unscathed, cementing his reputation for bravery.

- Harrison's inaugural address was so long that it took nearly two hours to deliver in freezing rain, a decision some believe weakened his health.
- He was the first president to die in office, prompting debates over whether Vice President Tyler should become "acting president" or fully assume the role, setting a precedent for future successions.

Ages at Death, Cause of Death, and Burial Locations

- **William Henry Harrison**: Died at age 68 on April 4, 1841, in Washington, D.C. The cause of death was pneumonia, likely contracted after his lengthy inauguration speech in cold weather, compounded by the era's limited medical treatments. He was buried at the William Henry Harrison Tomb State Memorial in North Bend, Ohio.
- **Anna Harrison**: Died at age 88 on February 25, 1864, in North Bend, Ohio. Her cause of death was natural causes, likely related to old age. She was buried alongside her husband at the William Henry Harrison Tomb State Memorial in North Bend, Ohio.

Conclusion

William Henry Harrison's life was a blend of military valor, political ambition, and tragic brevity. His rise from a Virginia aristocrat to a frontier hero and president showcased his adaptability and dedication to public service. Though his presidency was cut short, his military victories, particularly at Tippecanoe, and his role in shaping the Whig Party left a lasting mark. Anna Harrison's quiet strength supported their family and legacy, despite never fulfilling the role of

First Lady. Harrison's story is a reminder of the fragility of life and the unpredictable nature of leadership, cementing his place as the president remembered for the shortest tenure in American history.

John Tyler: The Accidental President

Introduction

John Tyler, the tenth President of the United States, is often remembered as the "Accidental President" due to his unprecedented ascension to the presidency following the death of William Henry Harrison in 1841. As the first vice president to assume the presidency upon a president's death, Tyler set a critical precedent for succession. His life, marked by a commitment to states' rights, political independence, and personal resilience, reflects the complexities of a nation on the brink of civil conflict. Born into Virginia's elite, Tyler's journey from a plantation upbringing to the White House was shaped by his steadfast principles, personal tragedies, and a turbulent political landscape. This summary explores Tyler's early life, family, rise to power, presidency, and legacy, highlighting his contributions, personality, and the unique anecdotes that define his story.

Early Life

John Tyler was born on March 29, 1790, in Charles City County, Virginia, into a prominent slaveholding family, embodying the Old Dominion's aristocracy. His upbringing on the Greenway plantation shaped his worldview, rooted in states' rights and Southern traditions.

- Educated at the College of William and Mary, graduating in 1807 at age 17, under the mentorship of Bishop James Madison, who served as a second father figure.

- Studied law under his father, a state judge, and later with Edmund Randolph, former U.S. Attorney General, gaining admission to the Virginia bar at 19 (despite being underage, as the judge overlooked his age).

- Inherited a significant number of slaves and properties, including Woodburn plantation, after his father's death in 1813, solidifying his status as a planter.

- Served briefly as a military captain during the War of 1812, leading a militia company but seeing no combat, which reinforced his regional loyalty.

- Early exposure to politics through his father, a former Virginia governor and friend of Thomas Jefferson, instilled a Jeffersonian commitment to limited federal power.

Family and Children

Tyler's personal life was defined by two marriages and a record-setting 15 children, the most of any U.S. president, reflecting his dedication to family despite financial strains.

- Married Letitia Christian in 1813, with whom he had eight children: Mary, Robert, John III, Letitia, Elizabeth, Alice, Tazewell, and an unnamed child who died in infancy. Letitia, an invalid by 1839 due to a stroke, died in 1842, becoming the first First Lady to die in the White House.

- Married Julia Gardiner in 1844, 30 years his junior, in a controversial White House wedding—the first for a sitting president. They had seven children: David Gardiner, John Alexander, Julia Gardiner, Lachlan, Lyon Gardiner, Robert Fitzwalter, and Pearl.

- Tyler's children spanned generations; his daughter Pearl died in 1947, and his grandson Harrison Ruffin Tyler lived until 2025, 235 years after Tyler's birth, due to Tyler and his son Lyon fathering children late in life.

- Financially burdened by his large family, Tyler often lived beyond his means, relying on nepotism (e.g., appointing his son John as press secretary) and plantation income, though debts mounted during the Civil War.

- Maintained close ties with his children, hosting lavish events like children's Christmas parties at the White House, despite political isolation.

Rise to Power

Tyler's political career began early, leveraging his family's influence and his own legal acumen to ascend through Virginia's political ranks, culminating in his unexpected presidency.

- Elected to the Virginia House of Delegates at 21 in 1811, serving until 1816, where he opposed federal overreach, including the Bank of the United States.

- Served in the U.S. House of Representatives from 1816 to 1821, voting against nationalist policies like the Missouri Compromise, reflecting his states' rights stance.

- Became Virginia's governor in 1825, elected by the state legislature, though the role lacked significant power, prompting his move to the U.S. Senate in 1827.

- Resigned from the Senate in 1836, refusing to comply with Virginia's legislature's instructions to reverse his vote against Andrew Jackson's policies, showcasing his principled independence.

- Nominated as vice president on the Whig ticket in 1840 with William Henry Harrison to attract Southern voters, winning with the slogan "Tippecanoe and Tyler Too." Harrison's death 31 days into his term thrust Tyler into the presidency on April 4, 1841.

Influences

Tyler's political philosophy and personal life were shaped by key figures and ideologies that reinforced his commitment to states' rights and constitutional strictness.

- His father, John Tyler Sr., a Jeffersonian advocate of states' rights, instilled a distrust of federal power and a belief in Virginia's sovereignty.

- Thomas Jefferson, a family friend, influenced Tyler's view of limited government and separation of church and state, evident in Tyler's support for religious freedom.

- Bishop James Madison at William and Mary provided intellectual guidance, fostering Tyler's legal and philosophical grounding in individual liberties.

- Southern planter culture, reliant on slavery, shaped Tyler's economic and social outlook, leading him to defend the institution and oppose federal interference.

- His opposition to Andrew Jackson's executive overreach, particularly on banking and tariffs, pushed Tyler toward the Whig Party, though he maintained Democratic roots.

Party Affiliation

Tyler's political alignment was fluid, reflecting his independent streak and commitment to principle over party loyalty, ultimately leaving him politically isolated.

- Began as a Democratic-Republican, aligning with Jeffersonian ideals of limited government and states' rights.

- Opposed Andrew Jackson's policies, particularly the removal of bank deposits, leading to a break with Democrats and alignment with the newly formed Whig Party in the 1830s.

- Ran as a Whig vice presidential candidate in 1840 but clashed with Whig leaders like Henry Clay over banking policies, resulting in his expulsion from the party in 1841.

- Became a "president without a party," briefly forming his own party in 1844 for a re-election bid, which he abandoned to support Democrat James K. Polk.

- Later aligned with Southern interests, supporting secession and serving in the Confederate House of Representatives, marking his final shift from Unionist to Confederate.

Presidency Accomplishments

Tyler's presidency, though marred by conflict, achieved notable successes, particularly in foreign policy and territorial expansion, despite his lack of party support.

- Signed the Preemption Act of 1841, allowing settlers to claim 160 acres of public land before sale, promoting westward expansion.

- Ended the Second Seminole War in Florida in 1842, reducing costly military engagements.

- Signed the Webster-Ashburton Treaty in 1842, resolving U.S.-British border disputes, including the Maine-Canada boundary, and stabilizing relations.

- Established the U.S. Weather Bureau, laying the foundation for modern meteorological services.

- Reorganized the U.S. Navy, enhancing its efficiency and structure.

- Signed the Tariff of 1842, a compromise that stabilized federal revenue while protecting Northern manufacturers.
- Secured the annexation of Texas in 1845 via joint resolution, a major expansionist achievement.
- Quelled Dorr's Rebellion in Rhode Island in 1842, reinforcing federal authority in domestic disputes.
- Negotiated the Treaty of Wanghia with China in 1844, opening Chinese ports to U.S. trade.

Contributions of the First Lady

Tyler's first two ladies, Letitia Christian and Julia Gardiner, had contrasting roles due to health and timing, with Julia leaving a more visible mark.

- **Letitia Christian Tyler:**

- Limited role due to a stroke, confined to the White House's second floor, making only one public appearance.
- Relied on daughter-in-law Priscilla Cooper as official hostess for White House events.
- Her death in 1842 was mourned nationally, marking a somber moment in Tyler's presidency.

- **Julia Gardiner Tyler:**

- Became First Lady in 1844 at age 24, the youngest in history, bringing vitality to the White House.
- Introduced "Hail to the Chief" as the president's ceremonial entrance music, a lasting tradition.

- Popularized the waltz and polka at White House functions, modernizing social events.

- Hosted elaborate balls and receptions, enhancing the White House's social prestige during Tyler's final months.

- Advocated for the annexation of Texas, influencing Tyler's expansionist agenda.

Positive and Negative Personality Traits and Their Impact on Presidency

Tyler's personality shaped his presidency, with his strengths and flaws contributing to both his achievements and challenges.

Positive Traits:

- Principled: Tyler's steadfast commitment to states' rights and constitutional strictness guided his vetoes of national bank bills, preserving his vision of limited federal power.

- Diplomatic: His courteous demeanor facilitated foreign policy successes like the Webster-Ashburton Treaty, calming tensions with Britain.

- Resilient: Despite political isolation, Tyler asserted full presidential authority, setting a succession precedent that strengthened the office.

Negative Traits:

- Stubborn: His refusal to compromise with Whigs on banking issues led to his expulsion from the party and cabinet resignations, crippling his administration's cohesion.

- Aloof: Perceived as elitist, Tyler struggled to connect with the populist base that elected Harrison, earning the derisive nickname "His Accidency."
- Inflexible: His rigid adherence to principles alienated both Whigs and Democrats, leaving him without a party and hindering legislative cooperation.
- **Impact**: Tyler's principled stance set important precedents but isolated him politically, limiting his domestic agenda. His diplomatic skills salvaged foreign policy wins, but his stubbornness fueled conflicts, including an impeachment attempt, marking his presidency as divisive.

Pets

Tyler's love for animals, particularly his horse, added a personal touch to his life, reflecting his Virginia planter lifestyle.

- Owned a favorite horse named "General," which he cherished and rode during his time at Sherwood Forest plantation.
- After the General's death, Tyler buried the horse on his plantation with a gravestone, a rare honor for a pet, highlighting his sentimental attachment.
- No specific records exist of White House pets, but Tyler's rural background suggests he likely kept dogs or other animals typical of plantation life.

Religious Persuasion

Tyler's religious beliefs, rooted in Jeffersonian ideals, emphasized the separation of church and state and personal liberty.

- Raised in the Episcopal Church, Tyler attended St. Paul's Episcopal Church in Richmond, where his funeral was held.

- Held Jeffersonian views on religious freedom, supporting a strict separation of church and state, as seen in his 1843 response to a Jewish leader's protest against military involvement in missionary activities.

- Believed religious expression should be private and free from government interference, aligning with his broader commitment to individual rights.

- His faith was understated but guided his moral framework, particularly in his efforts to preserve the Union before embracing secession.

Anecdotal Information

Tyler's life is punctuated by unique stories that reveal his character, quirks, and historical significance.

- Admitted to the Virginia bar at 19 by a judge who overlooked his underage status, showcasing his precocious legal talent and charm.

- Nicknamed "His Accidency" by detractors, a jab at his unelected rise to the presidency, which he countered by asserting full presidential authority.

- Proposed to Julia Gardiner shortly after Letitia's death, sparking a scandal due to their 30-year age gap and her youth, yet their marriage thrived.

- Hosted children's Christmas parties at the White House, reflecting his warmth as a father despite political turmoil.

- Named his plantation "Sherwood Forest," likening himself to Robin Hood after his Whig expulsion, a playful nod to his political outlaw status.

- His death in 1862 went unacknowledged by Washington due to his Confederate allegiance, a stark contrast to his earlier Unionist efforts.

Ages at Death, Causes of Death, and Burial Location

Tyler's final years were marked by declining health and political controversy, culminating in a death that reflected his divided loyalties.

- Died on January 18, 1862, at age 71 in Richmond, Virginia, likely from a stroke after suffering chills, dizziness, and suffocation.

- Letitia Christian Tyler died on September 10, 1842, at age 51, likely from complications of her stroke, in the White House.

- Julia Gardiner Tyler died on July 10, 1889, at age 69, in Richmond, Virginia, of unknown causes.

- Tyler was buried at Hollywood Cemetery in Richmond, alongside presidents James Monroe and Confederate leader Jefferson Davis, following a grand Confederate state funeral despite his request for simplicity.

- His death was not officially recognized in Washington due to his Confederate service, underscoring his controversial legacy.

Conclusion

John Tyler's life and presidency embody the contradictions of a nation grappling with its identity. As the "Accidental President," he navigated uncharted constitutional territory, setting a precedent for presidential succession that endures today. His commitment to states' rights, shaped by his Virginia roots and Jeffersonian influences, drove both his achievements—like the annexation of Texas—and his failures, including his estrangement from major parties. Tyler's personal life, marked by two marriages, 15 children, and a beloved horse named General, reveals a man of warmth and resilience amid political isolation. His presidency, though stormy, left lasting contributions to U.S. expansion and governance. Tyler's legacy, overshadowed by his Confederate allegiance, remains a complex chapter in American history, reflecting the challenges of leadership in a divided era.

James K. Polk: Architect of American Expansion

Description

James K. Polk, the 11th President of the United States, was a driven, methodical leader who significantly expanded American territory. Known for his work ethic and commitment to Manifest Destiny, Polk's presidency was marked by bold decisions and lasting impact, though his reserved demeanor and health issues shaped a complex legacy.

Introduction

James Knox Polk served as president from 1845 to 1849, a pivotal era in American history. His single term achieved major territorial gains, including the annexation of Texas and the acquisition of vast western lands. A staunch Jacksonian Democrat, Polk's focus on clear goals and efficient governance earned him the nickname "Young

Hickory." His presidency remains a study in ambition, achievement, and the costs of relentless dedication.

Early Life

Born November 2, 1795, in Pineville, North Carolina, Polk grew up in a frontier setting. His family moved to Tennessee when he was 11, shaping his rugged, determined character. Despite chronic health issues, including gallstones, he pursued education rigorously.

- Attended a Presbyterian academy and later the University of North Carolina, graduating with honors in 1818.

- Studied law under Felix Grundy, a prominent Nashville attorney, and was admitted to the Tennessee bar in 1820.

- Early exposure to politics came through his father, Samuel Polk, a surveyor and land speculator with Democratic-Republican ties.

Family

Polk's family was rooted in Scots-Irish heritage, with a strong work ethic and modest means. His parents, Samuel and Jane Knox Polk, instilled discipline and Presbyterian values. He was the eldest of 10 children, bearing early responsibilities.

- Married Sarah Childress in 1824, a partnership that bolstered his political career.

- No siblings played prominent public roles, but family ties in Tennessee aided his early political network.

Children

James and Sarah Polk had no children, a personal sorrow that directed their energies toward public life. Sarah's role as a political confidante filled the void, and their childlessness allowed her to focus on White House duties and Polk's agenda.

Rise to Power

Polk's political ascent was steady, built on legal acumen, oratory skills, and loyalty to Democratic principles. His career began in Tennessee, where he aligned with Andrew Jackson's faction.

- Served as a Tennessee state legislator (1823–1825), advocating for land reform.
- Elected to the U.S. House of Representatives (1825–1839), becoming Speaker of the House (1835–1839).
- Governor of Tennessee (1839–1841), though defeated in re-election bids.
- Emerged as a dark-horse Democratic presidential candidate in 1844, unifying party factions with his expansionist platform.

Influences

Polk's worldview was shaped by key figures and ideologies.

- Andrew Jackson, his mentor, instilled populist and democratic ideals.
- Manifest Destiny, the belief in America's divine right to expand, drove his territorial ambitions.

- Presbyterian upbringing emphasized duty and moral discipline, reflected in his governance style.
- Legal training under Felix Grundy honed his analytical approach to policy.

Party Affiliation

Polk was a lifelong Democrat, committed to the party's Jacksonian principles of limited government, states' rights, and economic opportunity for the common man. He opposed Whig policies like a national bank and high tariffs, advocating for agrarian interests and territorial expansion.

Presidency

Polk's presidency (1845–1849) was defined by a clear agenda: expand U.S. territory, reduce tariffs, establish an independent treasury, and settle the Oregon boundary. His administration faced challenges, including the Mexican-American War and sectional tensions over slavery.

- Annexed Texas in 1845, fulfilling a campaign promise.
- Negotiated the Oregon Treaty (1846) with Britain, securing the Pacific Northwest at the 49th parallel.
- Led the Mexican-American War (1846–1848), resulting in the Treaty of Guadalupe Hidalgo, which added California, Nevada, Utah, and parts of four other states.
- Established the Independent Treasury System (1846), ensuring federal funds were managed independently of private banks.
- Reduced tariffs through the Walker Tariff (1846), promoting free trade.

- Vetoed internal improvement bills, adhering to strict constitutionalism.

Accomplishments

- Secured the largest territorial expansion in U.S. history via the Mexican Cession and the Oregon Treaty.
- Strengthened the economy with low tariffs and the Independent Treasury.
- Fulfilled all major campaign promises, a rare feat for a president.
- Established the U.S. Naval Academy and Smithsonian Institution (as part of broader federal initiatives).
- Issued the first U.S. postage stamps, modernizing communication.

First Lady's Contributions

- Hosted bipartisan social events, fostering political dialogue despite Polk's reserved nature.
- Managed White House operations efficiently, banning dancing and alcohol to reflect their Presbyterian values.
- Served as Polk's closest advisor, editing speeches and shaping public image.
- Promoted cultural refinement, introducing gas lighting to the White House.
- Supported charitable causes, enhancing the first lady's role as a public figure.

Positive Traits

Polk's strengths shaped a focused, effective presidency.

- Determination: Relentlessly pursued goals, achieving his entire platform.
- Work ethic: Worked long hours, personally overseeing policy details.
- Strategic vision: Unified Democrats and navigated complex foreign and domestic issues. These traits drove territorial and economic successes but strained his health.

Negative Traits

Polk's flaws impacted his leadership and legacy.

- Aloofness: His reserved demeanor limited personal alliances, isolating him politically.
- Micromanagement: Overwork led to burnout and health decline.
- Ambiguity on slavery: Avoided addressing the issue directly, exacerbating sectional tensions. These traits hindered broader consensus and left unresolved conflicts for successors.

Pets

The Polks had no pets in the White House, reflecting their austere lifestyle. Their focus remained on governance and social duties, with little room for personal indulgences like animals.

Religious Persuasion

Polk was raised Presbyterian, adhering to its emphasis on discipline and morality. He attended services regularly but was not overtly devout. Late in life, he was baptized Methodist, influenced by Sarah's faith.

- Sarah Childress Polk was a devout Presbyterian, shaping their household's moral tone.
- Their faith guided White House policies, like banning alcohol and dancing.

Interesting Anecdotes

- Polk was the first president photographed in office, capturing a stern, formal image.
- He kept a diary, offering rare insight into his presidency and personal frustrations.
- During the Mexican-American War, he personally reviewed military dispatches, acting as his own commander-in-chief.
- Sarah banned music and dancing at White House events, earning the nickname "Sahara Sarah" from critics.

Age at Death

- James Polk died at 53 on June 15, 1849, three months after leaving office.
- Sarah Childress Polk died at 87 on August 14, 1891.

Cause of Death

- James Polk likely died of cholera, contracted in New Orleans during a post-presidency tour. His weakened health from overwork exacerbated his condition.

- Sarah Polk died of natural causes, likely old age, at her Nashville home.

Burial Location

- James Polk was buried on the grounds of Polk Place, his Nashville home, and later reinterred at the Tennessee State Capitol grounds.

- Sarah Polk was buried beside him at the Tennessee State Capitol grounds.

Conclusion

James K. Polk's presidency was a transformative chapter in American history, defined by unprecedented territorial expansion and economic reform. His relentless pursuit of Manifest Destiny doubled the nation's size, but his reserved nature and failure to address slavery sowed seeds of future conflict. Sarah Polk's contributions as a savvy first lady amplified his administration's impact. Though his life was cut short, Polk's legacy as the architect of American expansion endures, a testament to his vision and the costs of unyielding ambition.

Zachary Taylor: Hero of the Mexican-American War

Introduction

Zachary Taylor, the 12th President of the United States, is most remembered as a hero of the Mexican-American War, earning the nickname "Old Rough and Ready" for his gritty leadership and battlefield valor. Born into a prominent Virginia family, Taylor's life was shaped by the frontier, military service, and the contentious issue of slavery in a rapidly expanding nation. His brief presidency from 1849 to

1850 was marked by efforts to preserve the Union amid growing sectional tensions, though his sudden death curtailed his impact. This summary explores Taylor's early life, family, rise to power, influences, political affiliation, presidency, the contributions of First Lady Margaret Mackall Smith Taylor, his personality traits, pets, religious beliefs, notable anecdotes, and details of his death and burial, offering a comprehensive view of a man whose military legacy overshadowed his political tenure.

Early Life

Zachary Taylor was born on November 24, 1784, in Orange County, Virginia, likely at Hare Forest Farm or Montebello, estates tied to his family's planter heritage. His parents, Richard Taylor, a Revolutionary War veteran and lieutenant colonel, and Sarah Dabney Strother, descended from prominent Virginia families. In 1785, the family relocated to Kentucky, settling on a plantation called Springfield near Louisville. Growing up on the frontier, Taylor's childhood was marked by a rugged environment and limited formal education.

- Raised in a log cabin before his family's wealth grew, he eventually lived in a substantial brick house.

- Received a rudimentary education from tutors, as no schools were nearby, resulting in lifelong struggles with handwriting, spelling, and grammar.

- Developed frontier skills in hunting, horsemanship, and musketry, which shaped his military aspirations.

- By 1800, his father owned 10,000 acres and 26 enslaved people, embedding Taylor in a slaveholding culture from youth.

Taylor's early years instilled a practical, self-reliant mindset and a desire for a military career, setting the stage for his future as a soldier and leader.

Family and Children

In 1810, Zachary Taylor married Margaret Mackall Smith, a member of a prominent Maryland family, whom he met while stationed in Louisville. The couple settled in Baton Rouge, Louisiana, and owned plantations in Louisiana and Mississippi, worked by enslaved laborers. They had six children, though tragedy struck early with the loss of two daughters.

- Ann Margaret Mackall (1811–1875): Married Robert Crooke Wood, a surgeon, and lived a relatively private life.

- Sarah Knox (1814–1835): Married Jefferson Davis, future Confederate president, but died of malaria three months after their 1835 wedding.

- Octavia Pannill (1816–1820): Died young, likely from illness.

- Margaret Smith (1819–1820): Also died in childhood, a common tragedy of the era.

- Mary Elizabeth (1824–1909): Married William Wallace Smith Bliss; later served as White House hostess for her father due to her mother's frail health.

- Richard (1826–1879): Served as a Confederate general during the Civil War, reflecting the family's divided legacy.

The Taylors' family life was nomadic, often moving between military outposts, with Margaret managing the household amid frequent relocations. The loss of two children and the eventual prominence of their surviving children, particularly Richard, tied the family to both personal hardship and historical significance.

Rise to Power

Taylor's ascent to prominence was rooted in his 40-year military career, beginning in 1808 when he was commissioned as a first lieutenant in the U.S. Army. His service spanned multiple conflicts, culminating in his celebrated role in the Mexican-American War, which propelled him to national fame and the presidency.

- Joined the army in 1808, assigned to Fort Pickering (modern-day Memphis), and quickly proved adept at frontier command.

- Gained recognition in the War of 1812, defending Fort Harrison against Native American attacks in 1812.

- Fought in the Black Hawk War (1832) and the Second Seminole War (1835–1842), earning the nickname "Old Rough and Ready" for his leadership at the Battle of Lake Okeechobee (1837), where he was promoted to brigadier general.

- Achieved national hero status during the Mexican-American War (1846–1848), leading victories at Palo

Alto, Resaca de la Palma, Monterrey, and Buena Vista despite being outnumbered.

- His military fame led the Whig Party to nominate him for president in 1848, despite his lack of political experience and vague political stance.

Taylor's military successes, particularly in the Mexican-American War, made him a household name, enabling his unlikely rise to the presidency as a war hero with broad appeal.

Influences

Taylor's life and decisions were shaped by his military background, frontier upbringing, and the socio-political environment of his time. Key influences included:

- **Military Mentors**: Early commanders and his father's Revolutionary War service instilled a sense of duty and discipline, shaping his leadership style.

- **Frontier Life**: Growing up in Kentucky's wilderness fostered resilience and practical problem-solving, evident in his unpretentious military demeanor.

- **Slaveholding Culture**: As a plantation owner, Taylor was entrenched in the Southern slave economy, yet his nationalism led him to oppose slavery's expansion, reflecting a complex interplay of personal and national interests.

- **Unionist Ideals**: Inspired by figures like George Washington, Taylor prioritized national unity, influencing his hardline stance against secession during his presidency.

- **Whig Party Dynamics:** Though politically independent, the Whigs' strategic use of his war hero status shaped his candidacy, even as he resisted strict party alignment.

These influences molded Taylor into a pragmatic, Union-focused leader whose military mindset often clashed with the political demands of his era.

Party Affiliation

Zachary Taylor identified as a political independent but ran as the Whig Party candidate in the 1848 presidential election. His affiliation was strategic, driven by his national popularity rather than ideological alignment.

- Never voted in a presidential election before 1848, citing a desire not to oppose a potential commander-in-chief.

- Described himself as a Jeffersonian-Democrat in principle but leaned toward Whig ideals, favoring a strong banking system and limited executive power.

- Appealed to Northern Whigs for his military record and Southern Whigs for his slaveholding background, allowing the party to present him as a unifying figure.

- Won the 1848 election against Democrat Lewis Cass and Free Soil candidate Martin Van Buren, securing 163 electoral votes to Cass's 127.

- Maintained distance from party leaders like Henry Clay, prioritizing national unity over partisan politics, which isolated him from Congress.

Taylor's loose affiliation with the Whigs and independent streak made him an unconventional candidate, complicating his relationship with party operatives during his presidency.

Presidency Accomplishments

Zachary Taylor's presidency, lasting from March 4, 1849, to July 9, 1850, was brief and dominated by the slavery debate. His accomplishments were limited but significant in shaping the path toward the Compromise of 1850.

- Encouraged California and New Mexico to draft constitutions and apply for statehood, bypassing territorial status to avoid slavery disputes, correctly anticipating that both would prohibit slavery.

- Signed the Clayton-Bulwer Treaty (1850) with Britain, ensuring neither nation would control a proposed Central American canal, fostering an Anglo-American alliance.

- Confronted Southern secession threats in February 1850, vowing to personally lead the army to enforce federal laws and preserve the Union.

- Addressed foreign policy challenges, including tensions with Venezuela over Cuba and Spain's arrests of Americans, while maintaining diplomatic stability.

Taylor's focus on Union preservation and statehood initiatives set the stage for the Compromise of 1850, though his death prevented him from seeing it enacted.

Contributions of the First Lady

Margaret Mackall Smith Taylor, the First Lady, led a private life, avoiding public appearances due to frail health and a preference for domesticity. Her contributions were subtle but vital to the Taylor household.

- Managed the family's plantations in Louisiana and Mississippi, overseeing operations and enslaved laborers during Zachary's frequent military absences.

- Provided emotional stability for Taylor, maintaining a cohesive family unit despite constant relocations and the loss of two children.

- Delegated White House hostess duties to her daughter, Mary Elizabeth Bliss, due to her own health issues, ensuring social functions continued smoothly.

- Supported Taylor's military career by accompanying him to various outposts, adapting to harsh frontier conditions.

Margaret's behind-the-scenes role as a supportive spouse and household manager allowed Taylor to focus on his military and presidential duties, though her public contributions were minimal.

Positive Personality Traits

Taylor's personality was shaped by his military background and frontier upbringing, contributing both strengths and challenges to his presidency.

- **Courageous**: His battlefield bravery, evident in hand-to-hand combat at Monterrey and Buena Vista, earned him the loyalty of troops and public admiration.

- **Direct and Frank**: Taylor's straightforward communication, praised by Ulysses S. Grant, fostered trust among soldiers and some political allies.

- **Unassuming**: His lack of pretension, reflected in his informal dress and willingness to share soldiers' hardships, made him relatable and earned him the "Old Rough and Ready" nickname.

- **Nationalist**: His unwavering commitment to the Union, threatening to personally combat secessionists, strengthened his resolve to maintain national unity.

These traits bolstered his reputation as a decisive leader, enhancing his ability to confront crises like secession threats, though they did not fully translate to political success.

Negative Personality Traits

Taylor's personality also presented challenges that hindered his presidency, particularly in navigating the political landscape.

- **Politically Inexperienced**: Lacking prior political involvement, Taylor struggled to engage with Congress or articulate clear policy, leading to isolation.

- **Stubborn**: His refusal to align closely with Whig leaders or compromise on issues like the Wilmot Proviso alienated key political figures.

- **Isolated**: His preference for independence over party loyalty limited his ability to build coalitions, weakening his legislative influence.

- **Limited Education**: Poor writing and communication skills, stemming from his rudimentary education, undermined his ability to project authority in political correspondence.

These traits contributed to his administration's limited legislative impact and strained relations with Congress, leaving him sidelined in key debates like the Compromise of 1850.

Pets

Taylor's affection for animals was evident in his ownership of a wartime horse, Old Whitey, which became a notable companion during his military and presidential years.

- Old Whitey, a white horse, accompanied Taylor through the Mexican-American War and was often seen grazing on the White House lawn during his presidency.

- Known for sitting calmly atop Old Whitey amidst gunfire, Taylor's bond with the horse symbolized his unflappable demeanor.

- The horse's presence at the White House added a touch of informality, endearing Taylor to the public as a relatable figure.

- No records indicate other pets, but Old Whitey's prominence reflects Taylor's frontier roots and military identity.

Old Whitey's grazing on the White House lawn became a minor but memorable aspect of Taylor's brief tenure, reinforcing his unpretentious image.

Religious Persuasion

Taylor's religious beliefs were private and not well-documented, reflecting his reserved nature and focus on military and practical matters.

- Raised in a family with Anglican roots, common among Virginia planters, but no evidence suggests a strong personal devotion to a specific denomination.

- Postponed his inauguration from March 4 to March 5, 1849, out of respect for the Sabbath, indicating some adherence to Christian traditions.

- Attended numerous funerals during his presidency, including those of James K. Polk and Dolley Madison, suggesting a respect for religious ceremonies.

- His pragmatic, military-focused life likely prioritized duty over overt religious expression, aligning with his independent character.

Taylor's understated faith influenced minor decisions, like his inauguration delay, but did not significantly shape his public or political actions.

Anecdotal Information

Taylor's life is peppered with stories that highlight his unpretentious nature and military grit, offering insights into his character.

- During the Battle of Buena Vista, Taylor reportedly sat atop Old Whitey, calmly eating a biscuit under heavy fire, earning admiration for his cool-headedness.

- Coined the term "First Lady" in his eulogy for Dolley Madison in 1849, a phrase that became standard for presidents' wives.

- His tattered straw hat and informal attire during battles led soldiers to initially mistake him for a common soldier, enhancing his relatability.

- Despite owning enslaved people, Taylor instructed his overseer to prioritize their health and distributed small sums to them at Christmas, reflecting a complex, paternalistic view of slavery.

These anecdotes paint Taylor as a straightforward, resilient figure whose military persona overshadowed his political role.

Ages at Death, Causes of Death, and Burial Location

Zachary Taylor's sudden death cut short his presidency, leaving a legacy tied more to his military heroics than his political achievements.

- **Age at Death**: Died at 65 on July 9, 1850.

- **Cause of Death**: Succumbed to a stomach ailment, likely gastroenteritis or cholera morbus, possibly contracted from contaminated water or food (cherries and iced milk) consumed at a July 4, 1850, Washington Monument event. A 1991 exhumation found no evidence of poisoning, confirming natural causes.

- **Burial Location**: Initially interred in the Congressional Cemetery's Public Vault in Washington, D.C., from July 13 to October 25, 1850. His remains were later moved to the Zachary Taylor National Cemetery in Louisville, Kentucky, established in the 1920s, where he and Margaret (died 1852) rest in a limestone mausoleum with a granite base and marble interior.

Margaret Taylor died in 1852 at age 63, likely of natural causes, and was buried alongside Zachary in the family cemetery, later designated a national cemetery.

Conclusion

Zachary Taylor's life, defined by his role as a Mexican-American War hero, reflects a blend of frontier grit, military valor, and a brief, turbulent presidency. His early years in Kentucky shaped a practical, unassuming character, while his 40-year military career catapulted him to national fame. Despite his slaveholding background, Taylor's commitment to the Union and opposition to slavery's expansion marked his presidency, though his inexperience and short tenure limited his impact. Margaret Taylor's quiet support stabilized his personal life, while his personality—courageous yet politically naive—both aided and hindered his leadership. His beloved horse, Old Whitey, and understated religious observance added humanizing elements to his legacy. Taylor's sudden death at 65, likely from gastroenteritis, left the Compromise of 1850 to his successor, but his resolve to preserve the Union foreshadowed the challenges of the coming Civil War. Buried in the Zachary Taylor National Cemetery, his legacy endures as a soldier-statesman whose brief presidency bridged a critical moment in American history.

Millard Fillmore: The Compromise President

Description

Millard Fillmore, the 13th President of the United States, is often remembered for his role in passing the Compromise of 1850, a series of bills aimed at balancing sectional tensions over slavery. His presidency, marked by efforts to preserve the Union, reflected his pragmatic yet uninspiring leadership. Fillmore's legacy is tied to his attempts to navigate a deeply divided nation, though his decisions remain controversial.

Introduction

Millard Fillmore served as president from 1850 to 1853, ascending to the office after the death of President Zachary Taylor. A relatively obscure figure, Fillmore was a self-made man who rose from humble beginnings to the highest office in the land. His presidency is most remembered for the Compromise of 1850, which temporarily eased tensions between slave and free states. While his efforts aimed to maintain national unity, his support for policies like the Fugitive Slave Act drew criticism. This summary explores Fillmore's life, from his early struggles to his political career, family life, and lasting impact.

Early Life

Millard Fillmore was born on January 7, 1800, in a log cabin in Cayuga County, New York, to Nathaniel Fillmore and Phoebe Millard. His early life was marked by poverty and limited opportunities.

- Grew up in a rural, frontier environment with minimal formal education.
- Worked as an apprentice clothier and on his family's farm, enduring harsh labor.
- Self-educated, teaching himself to read and write with borrowed books.
- At age 15, he attended a local academy briefly, where he met his future wife, Abigail Powers.
- Moved to Buffalo, New York, in his early 20s, where he studied law and began his career.

Family

Fillmore's family life was stable but marked by personal tragedy, with his first wife's death shortly after his presidency.

- Married Abigail Powers in 1826, a teacher who encouraged his self-education.

- After Abigail's death in 1853, he married Caroline McIntosh, a wealthy widow, in 1858.
- Maintained close ties with his siblings, particularly his sister, Julia, who lived near him in Buffalo.
- His family provided emotional support during his political career, though his demanding schedule often kept him away.

Children

Fillmore had two children with his first wife, Abigail, who played significant roles in his personal life.

- Millard Powers Fillmore (1828–1889): Known as "Powers," he became a lawyer and served as his father's private secretary during the presidency.
- Mary Abigail Fillmore (1832–1854): Known as "Abbie," she was musically talented, often performing at White House events; she tragically died of cholera at age 22.
- Both children were deeply affected by their mother's death, and Fillmore remained devoted to their upbringing.

Rise to Power

Fillmore's political ascent was steady, driven by his work ethic and connections in New York politics.

- Admitted to the bar in 1823 and established a successful law practice in Buffalo.
- Elected to the New York State Assembly (1829–1831) as a member of the Anti-Masonic Party.

- Served in the U.S. House of Representatives (1833–1835, 1837–1843), aligning with the Whig Party.
- Became a prominent Whig leader, known for supporting economic development and tariffs.
- Elected Vice President in 1848 under Zachary Taylor; assumed presidency on July 9, 1850, after Taylor's death.

Influences

Fillmore's worldview was shaped by his modest upbringing and key figures in his life.

- Abigail Powers, his first wife, instilled a love of learning and intellectual discipline.
- Influenced by Whig leaders like Henry Clay, who championed compromise and economic growth.
- His frontier background fostered a belief in self-reliance and pragmatism.
- Exposure to Anti-Masonic ideals early in his career shaped his distrust of secret societies.
- Political mentor Thurlow Weed, a New York Whig leader, guided his early career.

Party Affiliation and Presidency

Fillmore was a Whig, a party favoring economic modernization and national unity, but his presidency highlighted his pragmatic approach over ideological purity.

- Served as president from July 9, 1850, to March 4, 1853.

- Pushed for the Compromise of 1850, a package of bills addressing slavery in new territories.
- Signed the Fugitive Slave Act, requiring the return of escaped slaves, which alienated abolitionists.
- Focused on economic policies, including infrastructure and trade expansion.
- Struggled to unify the fracturing Whig Party, leading to its decline and his failure to secure the 1852 nomination.

Accomplishments

Fillmore's presidency had notable achievements, though overshadowed by controversy.

- Successfully passed the Compromise of 1850, delaying Civil War tensions for a decade.
- Opened Japan to Western trade through Commodore Matthew Perry's expedition (results seen after his term).
- Supported the expansion of the Library of Congress, enhancing its national significance.
- Promoted infrastructure projects, including railroads and canals, to boost commerce.
- Maintained neutrality in foreign affairs, avoiding conflicts during a turbulent period.

Contributions of the First Lady

Abigail Fillmore served as First Lady from 1850 to 1853, leaving a lasting mark despite poor health.

- Established the first permanent White House library, a significant cultural contribution.

- Hosted intellectual gatherings, elevating the social prestige of the White House.

- Advised Fillmore privately on political matters, leveraging her education and insight.

- Promoted music and arts, with her daughter Abbie performing at events.

- Advocated for women's education, reflecting her background as a teacher.

Positive Personality Traits

Fillmore's positive traits shaped his ability to navigate complex political challenges.

- Pragmatic: Sought practical solutions to national issues, as seen in the Compromise of 1850.

- Diligent: His self-education and legal career reflected a strong work ethic.

- Diplomatic: Maintained calm leadership during sectional crises.

- Loyal: Remained committed to the Whig Party despite internal divisions. These traits helped him stabilize the nation temporarily, though his compromises were divisive.

Negative Personality Traits

Fillmore's flaws hindered his effectiveness and legacy.

- Indecisive: Often relied on advisors, leading to inconsistent policies.

- Overly cautious: Avoided bold action, which frustrated reformers and abolitionists.

- Lack of charisma: Failed to inspire the public or unify his party. These traits contributed to his inability to secure the 1852 Whig nomination and his fading political influence.

Pets

Fillmore's household included pets, though records are limited.

- Kept dogs at his Buffalo home, though specific names or breeds are undocumented.

- No evidence of pets in the White House, likely due to Abigail's health and formal duties.

Religious Persuasion

Fillmore's faith influenced his personal values but was not a dominant public feature.

- Raised in a Methodist family, but later joined the Unitarian Church in Buffalo.

- Unitarian beliefs emphasized reason and morality, aligning with his pragmatic outlook.

- Rarely emphasized religion in politics, focusing instead on national unity.

- Attended church regularly with Abigail, who shared his Unitarian faith.

Interesting Anecdotes

Fillmore's life included unique moments that reveal his character and era.

- As a young man, he walked miles to borrow books, reflecting his hunger for knowledge.
- During his presidency, he personally helped fight a fire at the Library of Congress in 1851.
- Refused an honorary degree from Oxford University, saying he hadn't earned it.
- After his presidency, he ran for president in 1856 with the nativist Know-Nothing Party, a controversial move.
- Enjoyed hosting buffalo hunts for visitors in Buffalo, a nod to his frontier roots.

Age at Death, Causes of Death, and Burial Locations

Fillmore and his first wife faced health challenges, with Abigail's death impacting him deeply.

- Millard Fillmore died on March 8, 1874, at age 74, in Buffalo, New York, from a stroke.
- Abigail Fillmore died on March 30, 1853, at age 55, in Washington, D.C., from pneumonia contracted at Franklin Pierce's inauguration.
- Both are buried at Forest Lawn Cemetery in Buffalo, New York.
- Caroline Fillmore, his second wife, died in 1881 and is also buried at Forest Lawn Cemetery.

Conclusion

Millard Fillmore's life reflects the challenges of leading a divided nation on the brink of civil war. His role in the Compromise of 1850 earned him the title of the "Compromise President," a legacy both praised for delaying conflict and criticized for appeasing slavery. From his humble beginnings to his pragmatic leadership, Fillmore's story is one of perseverance and moderation. His presidency, though not transformative, navigated a critical moment in American history. Abigail's contributions, particularly the White House library, added cultural depth to his tenure. While his caution and lack of charisma limited his impact, Fillmore's efforts to preserve the Union remain a defining, if contested, part of his legacy.

Franklin Pierce: Architect of the Kansas-Nebraska Act

Franklin Pierce, the 14th President of the United States, served from 1853 to 1857. Known for his role in passing the Kansas-Nebraska Act, his presidency was marked by escalating tensions over slavery that set the stage for the Civil War. A New Englander with a legal background, Pierce's charm and indecision shaped a turbulent era.

Introduction

Franklin Pierce's presidency is a study in contradiction: a charismatic leader whose well-intentioned policies deepened national divisions. Born in New Hampshire, Pierce rose through state and national politics as a Democrat, advocating for expansionist policies. His tenure is most remembered for the Kansas-Nebraska Act, which ignited violent conflict over slavery's expansion. This summary explores his life, family, political career, personality, and legacy, alongside the contributions of his wife, Jane Pierce, and lesser-known aspects like his pets and religious views.

Early Life

Franklin Pierce was born on November 23, 1804, in Hillsborough, New Hampshire, to Benjamin Pierce, a Revolutionary War veteran and two-term governor, and Anna Kendrick. Raised in a log cabin, Pierce grew up in a politically engaged household.

- Educated at local schools, he later attended Hancock Academy and Bowdoin College, graduating in 1824.
- At Bowdoin, he befriended Nathaniel Hawthorne, a lifelong influence.
- Studied law under Levi Woodbury, admitted to the bar in 1827, and began practicing in Hillsborough.
- Early exposure to politics through his father shaped his Democratic leanings.

Family

Pierce married Jane Means Appleton in 1834, a union that brought both devotion and tragedy. Jane, from a prominent but frail family, disliked public life, and their marriage was strained by loss and her poor health.

- Parents: Benjamin Pierce and Anna Kendrick; Jane's parents were Jesse Appleton, a Bowdoin president, and Elizabeth Means.
- Siblings: Pierce had seven siblings, including brothers who served in the military.

- Jane's reserved nature clashed with Pierce's sociable personality, creating tension during his political career.

Children

The Pierces had three sons, all of whom died young, profoundly impacting their lives.

- Franklin Jr. (1836): Died in infancy.
- Frank Robert (1839–1843): Died of typhus at age four.
- Benjamin (1841–1853): Died in a train accident at age 11, shortly before Pierce's inauguration, devastating Jane and haunting Pierce's presidency.
- The loss of Benjamin deepened Jane's depression and withdrawal from public duties.

Rise to Power

Pierce's political ascent was swift, driven by his charm, legal acumen, and Democratic loyalty.

- Elected to the New Hampshire House of Representatives at 24, becoming speaker by 1831.
- Served in the U.S. House (1833–1837) and Senate (1837–1842), resigning to focus on family and law.
- Gained prominence as a "dark horse" candidate at the 1852 Democratic National Convention, securing the nomination after 49 ballots.
- Defeated the Whig candidate Winfield Scott in the 1852 election, winning 27 of 31 states.

Influences

Pierce's political outlook was shaped by key figures and ideologies.

- His father, Benjamin, instilled a commitment to public service and Democratic principles.
- Levi Woodbury, his law mentor, introduced him to Jacksonian democracy.
- Friendship with Nathaniel Hawthorne provided intellectual and emotional support.
- Pierce admired Andrew Jackson's expansionist and populist policies, influencing his support for Manifest Destiny.

Party Affiliation and Presidency

Pierce, a lifelong Democrat, championed states' rights and territorial expansion. His presidency (1853–1857) was defined by efforts to maintain national unity amid growing sectional strife.

- Supported the Kansas-Nebraska Act (1854), allowing territories to decide on slavery, leading to "Bleeding Kansas" violence.
- Pursued expansionist policies, including the Gadsden Purchase (1853) and attempts to acquire Cuba.
- Enforced the Fugitive Slave Act, alienating Northern abolitionists.
- Struggled to unify a fractured Democratic Party, losing support by his term's end.

Accomplishments

Pierce's presidency had notable achievements, though overshadowed by controversy.

- Secured the Gadsden Purchase, adding southern territory for a transcontinental railroad.
- Negotiated trade agreements, including the Treaty of Kanagawa (1854) with Japan, opening ports to U.S. commerce.
- Reduced the national debt through fiscal discipline.
- Reformed the civil service, improving government efficiency.
- Strengthened U.S. naval and military capabilities.

Contributions of the First Lady

Jane Pierce, known as the "Shadow in the White House," was reclusive due to grief and poor health, limiting her public role.

- Advocated for temperance, reflecting her strict moral upbringing.
- Hosted limited White House events, maintaining decorum despite personal struggles.
- Supported charitable causes quietly, focusing on aid for the poor.
- Influenced Pierce's private decisions, though her withdrawal left him without a strong partner in public life.

Positive Personality Traits

Pierce's strengths shaped his early success but were less effective in the presidency.

- Charismatic and eloquent, he won allies easily in political circles.
- Loyal to friends and party, fostering strong networks.
- Empathetic, particularly in personal interactions, earning devotion from peers like Hawthorne.
- These traits helped him secure the presidency but were insufficient to navigate national crises.

Negative Personality Traits

Pierce's weaknesses contributed to his presidency's failures.

- Indecisive under pressure, he deferred to advisors like Jefferson Davis, leading to inconsistent policies.
- Overly conciliatory, his attempts to appease both North and South alienated both.
- Struggled with alcohol, particularly after personal tragedies, undermining his leadership.
- These traits exacerbated divisions, as his vacillation fueled distrust among factions.

Pets

Pierce's household included pets, though less documented than those of other presidents.

- Owned several horses, reflecting his love for riding and rural life.
- Likely had dogs, common in New Hampshire households, though specific names or breeds are unrecorded.
- Jane's delicate health limited pet involvement in White House life.

Religious Persuasion

Pierce's faith was private but significant.

- Raised in a Congregationalist family, he leaned toward Episcopalian practices after marriage.
- Jane's devout Calvinist beliefs influenced their household, emphasizing moral rigor.
- Pierce attended church regularly but avoided public displays of piety, aligning with his reserved approach to personal matters.

Interesting Anecdotes

Pierce's life is dotted with compelling stories.

- As a young lawyer, he once defended a client by riding 70 miles overnight to secure a witness.
- During the Mexican-American War, he fainted from a horse fall but returned to lead his troops, earning respect.
- Nathaniel Hawthorne wrote Pierce's campaign biography, a rare instance of a literary giant aiding a candidate.

- After Benjamin's death, Jane believed it was divine punishment, retreating into mourning during Pierce's term.

Age at Death, Causes of Death, and Burial Locations

Pierce and Jane faced health challenges, and their deaths reflected their struggles.

- Franklin Pierce died at 64 on October 8, 1869, in Concord, New Hampshire, from cirrhosis of the liver, likely tied to chronic alcoholism. Buried at Old North Cemetery, Concord.
- Jane Pierce died at 57 on December 2, 1863, in Andover, Massachusetts, from tuberculosis. Buried at Old North Cemetery, Concord, beside Franklin.

Conclusion

Franklin Pierce's life was a blend of ambition, tragedy, and miscalculation. A gifted politician with a knack for winning hearts, he was ill-equipped to bridge a nation fracturing over slavery. His Kansas-Nebraska Act, meant to preserve unity, instead accelerated conflict, cementing his legacy as a well-meaning but flawed leader. Jane's limited role as First Lady reflected their personal grief, which shadowed his presidency. Despite accomplishments like the Gadsden Purchase and trade expansion, Pierce's indecision and inability to manage sectional tensions define his place in history. His story is a poignant reminder of leadership's limits in the face of irreconcilable divides.

James Buchanan: The President Who Failed to Avert Civil War

Introduction

James Buchanan, the 15th President of the United States, served from 1857 to 1861, a period marked by escalating tensions over slavery that pushed the nation toward civil war. His presidency is often criticized for its inaction and inability to unify a divided country. A career politician with a legal background, Buchanan's cautious and legalistic approach to governance failed to address the growing sectional crisis, earning him a reputation as one of America's least effective presidents. His life, however, reflects the complexities of a man navigating a turbulent era, shaped by personal loss, ambition, and a commitment to the Union that ultimately fell short.

Early Life

James Buchanan was born on April 23, 1791, in Cove Gap, Pennsylvania, to a prosperous merchant family. His early years were defined by a strong work ethic and access to education, which set the stage for his political career.

- Grew up in a log cabin in rural Pennsylvania, the second of eleven children.

- Attended Old Stone Academy and later Dickinson College, graduating in 1809 despite disciplinary issues.

- Studied law in Lancaster, Pennsylvania, and was admitted to the bar in 1812, quickly establishing a successful legal practice.

- Served briefly in the War of 1812 as a volunteer in a militia, though he saw no combat.

Family

Buchanan's personal life was marked by tragedy and solitude, as he remained the only bachelor president in U.S. history.

- Parents: James Buchanan Sr., a merchant and farmer, and Elizabeth Speer, a well-educated woman who influenced his intellectual development.

- Never married, possibly due to the death of his fiancée, Ann Coleman, in 1819, who broke off their engagement amid rumors of Buchanan's infidelity or financial motives. Her sudden death, possibly by suicide, deeply affected him.

- Maintained close ties with his siblings, particularly his sister, Harriet, and took responsibility for supporting his nieces and nephews after their parents' deaths.

- Lived with his niece, Harriet Lane, who served as his White House hostess during his presidency.

Children

- Buchanan had no children of his own, as he never married.

- He acted as a guardian to several nieces and nephews, including Harriet Lane, whom he treated as a daughter and who played a significant role in his public life.

Rise to Power

Buchanan's political career spanned over four decades, marked by steady ambition and a knack for diplomacy, though he often avoided taking bold stands.

- Began in the Pennsylvania House of Representatives (1814–1816), aligning with the Federalist Party initially.

- Served in the U.S. House of Representatives (1821–1831) as a Democratic-Republican, later a Democrat.

- Appointed Minister to Russia (1832–1833) under Andrew Jackson, where he negotiated a trade treaty.

- Served as U.S. Senator from Pennsylvania (1834–1845), chairing the Foreign Relations Committee.

- Became Secretary of State under President James K. Polk (1845–1849), negotiating the Oregon Treaty with Britain.

- Served as Minister to the United Kingdom (1853–1856) under Franklin Pierce, helping draft the Ostend Manifesto, a controversial plan to acquire Cuba.

- Nominated as the Democratic candidate for president in 1856, he was seen as a "safe" choice due to his experience and absence during domestic controversies like the Kansas-Nebraska Act.

Influences

Buchanan's worldview was shaped by his legal training, devotion to the Constitution, and a belief in limited federal power, which often constrained his actions during crises.

- Influenced by Andrew Jackson's strong Unionism, though he lacked Jackson's decisiveness.
- His Pennsylvania upbringing exposed him to both Northern and Southern perspectives on slavery, leading to his preference for compromise.
- Mentored by older Federalists and Democrats, he adopted a cautious, legalistic approach to politics.
- Ann Coleman's death left him emotionally reserved, possibly contributing to his focus on career over personal life.

Party Affiliation and Presidency

Buchanan was a lifelong Democrat, committed to the party's principles of states' rights and limited government, but his presidency struggled to navigate the slavery issue.

- Elected president in 1856, defeating Republican John C. Frémont and Know-Nothing Millard Fillmore, with strong Southern support.

- Presidency (1857–1861) was dominated by the slavery debate, particularly the Kansas Territory's violent conflicts ("Bleeding Kansas").
- Supported the pro-slavery Lecompton Constitution in Kansas, alienating Northern Democrats and escalating tensions.
- Backed the Dred Scott decision (1857), believing it would settle the slavery question, but it inflamed Northern opposition.
- Failed to decisively address the Panic of 1857, a financial crisis that deepened economic divides.
- Attempted to distract from domestic issues with foreign policy, including failed efforts to annex Cuba and interventions in Latin America.
- His inaction as Southern states began seceding in late 1860 left the Union vulnerable, deferring major decisions to Abraham Lincoln.

Accomplishments

Buchanan's presidency is often viewed as a failure, but he had some achievements, mostly in foreign policy.

- Negotiated the Treaty of Washington (1860), settling disputes with Britain over Central America.
- Strengthened U.S. influence in the Pacific by establishing trade relations with Japan.
- Maintained neutrality during European conflicts, avoiding entanglement in foreign wars.

- Oversaw the Utah War (1857–1858), resolving tensions with Mormon settlers through negotiation rather than prolonged conflict.

Contributions of the First Lady

As Buchanan was unmarried, his niece, Harriet Lane, served as White House hostess, earning praise for her charm and social influence.

- Acted as an unofficial First Lady, managing White House social events with sophistication.
- Promoted American art and music, hosting cultural events that elevated Washington's social scene.
- Advocated for Native American rights, a rare cause for the time, though her efforts had a limited impact.
- Used her position to support charitable causes, including hospitals and orphanages.
- Set a precedent for future First Ladies by balancing diplomacy and social leadership.

Positive and Negative Personality Traits

Buchanan's personality shaped his presidency, often to its detriment.

- **Positive Traits**: Diligent, detail-oriented, and experienced in diplomacy, which aided his foreign policy efforts. His loyalty to the Constitution reflected a deep commitment to the Union.
- **Negative Traits**: Indecisive, overly cautious, and prone to deferring to advisors, particularly pro-Southern ones,

which weakened his leadership. His legalistic mindset prevented bold action, and his aloof demeanor alienated allies.

- **Impact on Presidency**: His caution led to paralysis during crises like Bleeding Kansas and secession, while his pro-Southern bias damaged his credibility in the North, undermining his ability to unify the nation.

Pets

Buchanan's household included a few notable pets, reflecting his personal side.

- Owned a Newfoundland dog named Lara, known for her gentle temperament, often mentioned in letters.
- Kept an eagle, a gift from supporters, which was housed on the White House grounds.
- Had a pair of pet mockingbirds, which he reportedly enjoyed for their singing.

Religious Persuasion

Buchanan's faith was private but consistent, aligning with his reserved personality.

- Raised in the Presbyterian Church, he remained a member throughout his life.
- Attended services regularly but avoided public displays of religiosity, reflecting his belief in separating personal faith from governance.

- His faith influenced his moral stance against war and his preference for peaceful resolutions, as seen in the Utah War.

Interesting Anecdotes

Buchanan's life included quirky and revealing moments that humanize his otherwise formal persona.

- During his engagement to Ann Coleman, rumors spread that he was more interested in her wealth than her affection, leading to their breakup and her tragic death, which haunted him lifelong.
- As Minister to Russia, he struggled with the cold climate and once humorously complained about wearing "seven layers of clothing" to survive St. Petersburg winters.
- Known for his fastidiousness, he once halted a cabinet meeting to correct a misspelled word in a document, showcasing his obsession with detail.
- At the White House, he reportedly allowed his pet eagle to wander the grounds, startling guests and staff.

Age at Death, Causes of Death, and Burial Locations

Buchanan and Harriet Lane's later years and final resting places reflect their distinct roles in history.

- **James Buchanan**: Died on June 1, 1868, at age 77, at his home, Wheatland, in Lancaster, Pennsylvania. Cause of death was respiratory failure, likely due to pneumonia. Buried at Woodward Hill Cemetery in Lancaster.

- **Harriet Lane**: Died on July 3, 1903, at age 73, in Narragansett, Rhode Island. Cause of death was cancer. Buried at Green Mount Cemetery in Baltimore, Maryland.

Conclusion

James Buchanan's life was a paradox of ambition and failure. A skilled diplomat and seasoned politician, he rose to the presidency with a wealth of experience but lacked the vision and decisiveness to address the nation's greatest crisis. His commitment to the Union was genuine, yet his legalistic caution and pro-Southern leanings exacerbated divisions, leaving a legacy of missed opportunities. Harriet Lane's grace as White House hostess provided a bright spot in his administration, but it could not offset his inability to avert the Civil War. Buchanan's story is a cautionary tale of leadership faltering in the face of monumental challenges, remembered most for his failure to preserve national unity on the eve of America's greatest conflict.

Abraham Lincoln: The Great Emancipator

Introduction

Abraham Lincoln, the 16th President of the United States, is one of the most revered figures in American history. Known as the Great Emancipator, he led the nation through its darkest period—the Civil War—while championing the abolition of slavery and preserving the Union. His humble beginnings, moral conviction, and eloquent leadership shaped a legacy that continues to inspire. This summary explores Lincoln's life, from his early years to his presidency, personal traits, family, and lasting impact.

Early Life

Born on February 12, 1809, in a log cabin in Hardin County, Kentucky, Abraham Lincoln grew up in poverty on the American frontier. His early life was marked by hardship, limited education, and a strong work ethic that shaped his character.

- Moved to Indiana in 1816, where he worked on the family farm and faced the loss of his mother, Nancy Hanks Lincoln, at age nine.

- Largely self-educated, Lincoln devoured books, developing a love for learning despite attending school for less than a year.

- Relocated to Illinois in 1830, where he worked various jobs, including rail-splitter, store clerk, and surveyor, laying the foundation for his public life.

Family

Lincoln's family life was both a source of strength and personal tragedy, influencing his emotional depth and resilience.

- Married Mary Todd on November 4, 1842, a well-educated woman from a prominent Kentucky family.

- His father, Thomas Lincoln, was a farmer and carpenter; his stepmother, Sarah Bush Johnston, encouraged his education.

- Lincoln maintained distant relationships with his father but was deeply affected by his mother's and sister's early deaths.

Children

Abraham and Mary Lincoln had four sons, though only one survived to adulthood, shaping their personal lives with both joy and profound grief.

- Robert Todd Lincoln (1843–1926): The only son to reach adulthood, became a lawyer and diplomat.

- Edward Baker Lincoln (1846–1850): Died young of likely tuberculosis, deeply affecting the Lincolns.
- William Wallace Lincoln (1850–1862): Died during Lincoln's presidency, devastating the family.
- Thomas "Tad" Lincoln (1853–1871): Died at 18, further compounding Mary's emotional struggles.

Rise to Power

Lincoln's ascent to the presidency was driven by his intellect, oratory, and commitment to principle, despite lacking formal education or elite status.

- Began his political career in 1834 as a member of the Illinois State Legislature, serving four terms as a Whig.
- Practiced law in Springfield, Illinois, earning a reputation for honesty and eloquence.
- Elected to the U.S. House of Representatives in 1846, where he opposed the Mexican-American War, gaining national attention.
- Rose to prominence in the 1850s through debates with Stephen Douglas, advocating against the expansion of slavery, which propelled him to the Republican nomination in 1860.
- Won the presidency in 1860 with 39.8% of the popular vote, leading to Southern secession and the Civil War.

Influences

Lincoln's worldview was shaped by a blend of personal experiences, intellectual curiosity, and moral conviction.

- Inspired by the Declaration of Independence and its principles of liberty and equality.

- Influenced by thinkers like Thomas Paine and John Stuart Mill, whose writings reinforced his belief in individual rights.

- His frontier upbringing instilled a deep empathy for the common man, guiding his policies.

- Political mentors like Henry Clay, a Whig leader, shaped his views on economic development and national unity.

Party Affiliation and Presidency

Lincoln's political career evolved from Whig to Republican, reflecting his commitment to anti-slavery principles and Union preservation.

- Initially a Whig, he joined the newly formed Republican Party in 1856, aligning with its anti-slavery platform.

- Elected president in 1860, serving from March 4, 1861, until his assassination in 1865.

- Led the Union through the Civil War, balancing military strategy, political alliances, and moral imperatives.

- Issued the Emancipation Proclamation in 1863, redefining the war as a fight for freedom.

- Delivered the Gettysburg Address in 1863, articulating the war's purpose and the nation's democratic ideals.

- Re-elected in 1864, advocating for Reconstruction and reconciliation with the South.

Accomplishments

Lincoln's presidency was defined by monumental achievements that reshaped the United States.

- Preserved the Union by leading the North to victory in the Civil War, preventing the nation's permanent division.
- Issued the Emancipation Proclamation (1863), freeing enslaved people in Confederate territories and paving the way for the 13th Amendment, which abolished slavery nationwide.
- Signed the Homestead Act (1862), granting land to settlers, promoting westward expansion.
- Established the U.S. Department of Agriculture and supported the Morrill Act (1862), creating land-grant colleges to advance education.
- Delivered the Gettysburg Address, redefining American values of equality and liberty.
- Laid the groundwork for Reconstruction, promoting a lenient plan to reintegrate Southern states.

Contributions of the First Lady

Mary Todd Lincoln, as First Lady, played a complex role, marked by both contributions and controversies.

- Hosted social events at the White House, fostering political alliances during the Civil War.
- Advocated for the welfare of Union soldiers, visiting hospitals and raising funds for their care.

- Supported African American causes, including aid for freedmen and contraband camps.
- Renovated the White House, aiming to restore its dignity, though her spending drew criticism.
- Provided emotional support to Lincoln, despite her own struggles with mental health and public scrutiny.

Positive and Negative Personality Traits

Lincoln's personality profoundly influenced his leadership, blending strengths with human flaws.

Positive Traits:

- Empathy: His compassion for others, rooted in his humble origins, guided his policies on slavery and Reconstruction.
- Eloquence: His speeches, like the Gettysburg Address, unified and inspired the nation.
- Resilience: He endured personal losses and political setbacks, maintaining focus on national unity.
- Honesty: Known as "Honest Abe," his integrity earned trust among allies and adversaries.

Negative Traits:

- Indecisiveness: Early in his presidency, he struggled with military appointments, frustrating advisors. Meanwhile, Lincoln's indecision was due to his reluctance to shed blood, a trait most would consider admirable, but it did cause some tensions in his administration.

- Melancholy: His bouts of depression sometimes affected his demeanor, though he channeled this into reflective leadership.
- These traits shaped a presidency that was both deeply human and transformative, balancing moral conviction with pragmatic governance.

Pets

The Lincoln family's pets brought warmth to the White House during turbulent times.

- Lincoln was fond of animals, particularly cats, and often played with kittens in the White House.
- Owned a dog named Fido, a yellow mongrel left behind in Springfield due to the move to Washington.
- Sons Willie and Tad had goats, Nanny and Nanko, which roamed the White House grounds.
- Kept a pet turkey named Jack, originally intended for Christmas dinner but spared by Tad's plea.

Religious Persuasion

Lincoln's religious beliefs were complex and evolved over time, reflecting a private spirituality rather than formal adherence.

- Raised in a Baptist family but never joined a church, skeptical of organized religion.
- Frequently referenced God in speeches, particularly during the Civil War, suggesting a belief in divine providence.

- Read the Bible regularly, drawing moral guidance, though he avoided dogmatic affiliations.
- His faith deepened during personal tragedies and the war, shaping his views on fate and justice.

Interesting Anecdotes

Lincoln's life was filled with stories that highlight his humor, humility, and humanity.

- As a young lawyer, he once walked miles to return a small sum of money he'd overcharged a client, earning the nickname "Honest Abe."
- During the Civil War, he frequently visited telegraph offices late at night, anxiously awaiting battlefield news.
- He once pardoned a soldier sentenced to death for sleeping on duty, reflecting his compassion for ordinary soldiers.
- Lincoln enjoyed telling humorous stories, often using them to defuse tense political situations or connect with audiences.

Age at Death, Causes of Death, and Burial Locations

Lincoln's life ended tragically, cutting short his vision for a reunited nation.

- **Abraham Lincoln**: Died at age 56 on April 15, 1865, after being shot by John Wilkes Booth at Ford's Theatre in Washington, D.C. He succumbed to the gunshot wound to the head the next morning at the Petersen House. Buried in Oak Ridge Cemetery, Springfield, Illinois.

- **Mary Todd Lincoln**: Died at age 63 on July 16, 1882, likely from a stroke, in Springfield, Illinois. Buried beside Abraham in Oak Ridge Cemetery.

Conclusion

Abraham Lincoln's legacy as the Great Emancipator endures through his leadership in preserving the Union and abolishing slavery. His journey from a frontier cabin to the White House reflects the American ideal of opportunity and resilience. Despite personal and political challenges, Lincoln's empathy, eloquence, and moral vision transformed the nation. His presidency not only ended slavery but also redefined the United States as a nation committed to equality and unity. Lincoln's life remains a testament to the power of perseverance and principle in shaping history.

Andrew Johnson: The Reconstruction President

Andrew Johnson, the 17th President of the United States, was a complex figure known for his staunch Unionism and controversial Reconstruction policies following the Civil War. A self-made man with humble origins, he rose from a tailor's apprentice to a key political leader, only to face impeachment during his presidency due to his lenient approach toward the South and clashes with Congress.

Introduction

Andrew Johnson served as president from April 15, 1865, to March 4, 1869, assuming office after Abraham Lincoln's assassination. His tenure was marked by efforts to restore the Union quickly, but his resistance to Radical Republican policies aimed at protecting freed slaves' rights led to significant political turmoil. Johnson's presidency is most remembered for his role in Reconstruction and surviving an impeachment trial by a single Senate vote.

Early Life

Born into poverty, Johnson's early years shaped his resilience and populist views.

- Born: December 29, 1808, in Raleigh, North Carolina.

- Parents: Jacob Johnson, a porter, and Mary "Polly" McDonough, a seamstress.

- Education: Largely self-taught, with no formal schooling; learned to read and write as a young apprentice.

- Occupation: Apprenticed as a tailor at age 10, later opening his own tailoring shop in Greeneville, Tennessee.

- Early Challenges: Orphaned at a young age after his father's death and his mother's remarriage, he relied on hard work to survive.

Family

Johnson's family life was anchored by his marriage to Eliza McCardle, who supported his political ambitions despite her frail health.

- Marriage: Married Eliza McCardle on May 17, 1827, in Greeneville, Tennessee.

- Eliza's Role: Taught Johnson writing and arithmetic, fostering his self-education.

- Family Dynamics: The couple faced financial struggles early on but built a stable home in Tennessee.

Children

Andrew and Eliza Johnson had five children, though their lives were often marked by tragedy.

- Martha Johnson (1828–1901): Acted as White House hostess due to her mother's ill health.

- Charles Johnson (1830–1863): A Union army surgeon, died after a fall from a horse during the Civil War.

- Mary Johnson (1832–1883): Married Daniel Stover; also helped with White House duties.

- Robert Johnson (1834–1869): Struggled with alcoholism, served as Johnson's secretary, and died by suicide.

- Andrew "Frank" Johnson Jr. (1852–1879): Youngest son, died of tuberculosis.

Rise to Power

Johnson's political career began locally and grew through his alignment with working-class interests.

- Early Politics: Elected alderman in Greeneville (1829), then mayor (1834).

- State Level: Served in the Tennessee House (1835–1837, 1839–1841) and Senate (1841–1843).

- National Stage: U.S. House of Representatives (1843–1853), Tennessee governor (1853–1857), U.S. Senator (1857–1862).

- Vice Presidency: Chosen as Lincoln's running mate in 1864 for his Union loyalty; became vice president in March 1865.
- Presidency: Assumed office after Lincoln's assassination on April 14, 1865.

Influences

Johnson's worldview was shaped by his modest beginnings and admiration for certain political figures.

- Key Figures: Admired Andrew Jackson for his populism and Unionism.
- Ideology: Believed in limited government, states' rights, and the Constitution as a guiding framework.
- Social Context: His tailor background instilled empathy for the working class, but limited his understanding of racial equality.

Party Affiliation

Johnson's political allegiance shifted during his career, reflecting his pragmatic approach.

- Early Career: Democrat, advocating for small farmers and laborers.
- Civil War Era: Joined the National Union Party (a coalition of Republicans and War Democrats) to support Lincoln's Union efforts.
- Presidency: Reverted to Democratic principles, clashing with Republican-dominated Congress.

Presidency

Johnson's presidency focused on rapid Reconstruction but alienated many with his leniency toward former Confederates.

- Reconstruction Policies: Issued pardons to ex-Confederates, allowed Southern states to rejoin the Union with minimal conditions.
- Vetoes: Opposed the Civil Rights Act of 1866 and Freedmen's Bureau bills, overridden by Congress.
- Impeachment: Violated the Tenure of Office Act by removing Secretary of War Edwin Stanton; impeached by the House in 1868, acquitted in the Senate by one vote.
- Foreign Policy: Oversaw the purchase of Alaska from Russia in 1867, known as "Seward's Folly."

Accomplishments

- Successfully restored the Union by readmitting the Southern states.
- Oversaw the purchase of Alaska, expanding U.S. territory.
- Maintained fiscal restraint, reducing national debt post-Civil War.

First Lady's Contributions

- Supported Johnson's political career through education and encouragement.

- Maintained family stability despite personal health challenges.
- Delegated White House hostess duties to daughters Martha and Mary due to illness.

Positive Traits

Johnson's strengths included determination and loyalty to the Union.

- Resilience: Overcame poverty through self-education and hard work.
- Unionism: Remained loyal to the Union as a Southern senator, earning Lincoln's trust.
- Effect: His commitment to Union restoration drove his Reconstruction policies, aiming for national unity.

Negative Traits

Johnson's flaws significantly hindered his presidency.

- Stubbornness: Refused to compromise with Congress on Reconstruction, escalating tensions.
- Racial Bias: Opposed equal rights for freed slaves, undermining civil rights efforts.
- Effect: His inflexibility and prejudices led to impeachment and weakened his legacy.

Pets

The Johnson family kept minimal pets, likely due to their focus on political and family challenges.

- Known Pets: No specific pets documented in White House records; family prioritized survival and duty over pet ownership.

Religious Persuasion

Johnson's faith was personal but not overtly influential in his public life.

- Affiliation: No formal church membership, though raised in a Christian household.
- Practices: Attended Methodist and Presbyterian services occasionally with Eliza.
- Impact: Religion played a minor role in his presidency, overshadowed by political conflicts.

Interesting Anecdotes

Johnson's life was filled with unique moments that highlighted his character.

- Tailor's Pride: Continued sewing clothes as a senator, symbolizing his humble roots.
- Drunken Speech: Appeared intoxicated at his vice-presidential inauguration in 1865, damaging his reputation.
- Impeachment Drama: His Senate trial captivated the nation, with crowds gathering for updates.

Ages at Death

- Andrew Johnson: Died at 66 on July 31, 1875.
- Eliza McCardle Johnson: Died at 65 on January 15, 1876.

Causes of Death

- Andrew Johnson: Suffered a stroke while visiting family in Carter County, Tennessee.
- Eliza McCardle Johnson: Died of tuberculosis, exacerbated by years of poor health.

Burial Locations

- Andrew Johnson: Buried at Andrew Johnson National Cemetery, Greeneville, Tennessee, wrapped in a U.S. flag with the Constitution under his head, per his request.
- Eliza McCardle Johnson: Buried beside Andrew at Andrew Johnson National Cemetery, Greeneville, Tennessee.

Conclusion

Andrew Johnson's presidency remains a pivotal yet contentious chapter in American history. His dedication to the Union and rapid Reconstruction efforts were overshadowed by his resistance to civil rights and near-removal from office. A man of humble origins, Johnson's stubbornness and racial biases limited his effectiveness, leaving a mixed legacy as the Reconstruction President. His and Eliza's lives reflect resilience amid personal and national turmoil, with their story preserved in Greeneville, Tennessee.

Ulysses S. Grant: Hero of the Civil War and Reconstruction

Introduction

Ulysses S. Grant, the 18th President of the United States, is best remembered as the Union general who secured victory in the Civil War and as a president who championed Reconstruction to rebuild a fractured nation. His life was marked by military triumphs, political challenges, and personal resilience, though his presidency faced criticism for corruption scandals. This summary explores his life, from humble beginnings to his lasting legacy.

Early Life

Born Hiram Ulysses Grant on April 27, 1822, in Point Pleasant, Ohio, Grant grew up in a modest family. His early years shaped his tenacity and unassuming nature.

- Raised in Georgetown, Ohio, in a small, rural community.

- Worked in his father's tannery, developing a strong work ethic but disliking the trade.
- Showed early skill with horses, becoming an expert equestrian.
- Attended local schools, described as quiet but diligent, with a knack for mathematics.

Family

Grant's family provided a foundation of support, though his relationships were sometimes strained by his career and financial struggles.

- Father: Jesse Root Grant, a tanner and merchant, pushed Grant toward education and self-reliance.
- Mother: Hannah Simpson Grant, reserved but influential, instilled stoic values.
- Siblings: Grant was the eldest of six, maintaining close ties with his family despite frequent moves.
- Later reconciled with his father after initial disapproval of his military career.

Children

Grant and his wife, Julia, had four children, whom he adored and prioritized despite his demanding career.

- Frederick Dent Grant (1850–1912): Became a soldier and diplomat, following his father's path.
- Ulysses S. Grant Jr. (1852–1929): Known as "Buck," pursued business ventures, some unsuccessful.

- Ellen Wrenshall Grant (1855–1922): Called "Nellie," her White House wedding was a national event.
- Jesse Root Grant (1858–1934): The youngest, later a businessman and political figure.

Rise to Power

Grant's ascent from obscurity to national hero was driven by his military prowess and determination during the Civil War.

- Appointed to West Point in 1839, graduating in 1843, ranked 21st of 39.
- Served in the Mexican-American War (1846–1848), earning praise for bravery under General Zachary Taylor.
- Resigned from the army in 1854 amid personal struggles, including rumored drinking issues.
- Struggled as a civilian, failing in farming and business ventures in Missouri and Illinois.
- Rejoined the army in 1861 as the Civil War began, quickly rising to prominence.
- Key victories at Shiloh, Vicksburg, and Chattanooga cemented his reputation.
- Appointed general-in-chief of the Union Army in 1864, leading to the Confederate surrender in 1865.
- National fame led to his nomination as the Republican candidate for president in 1868.

Influences

Grant's life and leadership were shaped by key figures and experiences.

- General Zachary Taylor, his Mexican-American War mentor, modeled straightforward leadership.
- Abraham Lincoln's trust in Grant's military judgment boosted his confidence and career.
- Julia Dent Grant, his wife, provided emotional stability and encouraged his ambitions.
- Civil War experiences instilled a belief in national unity and equal rights, guiding his presidency.

Party Affiliation

Grant aligned with the Republican Party, which dominated post-Civil War politics.

- Joined the Republicans in the 1860s, drawn to their Unionist and anti-slavery stance.
- Ran as a Republican in 1868 and 1872, winning both elections decisively.
- Supported Radical Republican policies during Reconstruction, advocating for African American rights.
- Later, he distanced himself from party factionalism, focusing on national unity.

Presidency

Grant served two terms as president from 1869 to 1877, a period defined by Reconstruction and economic challenges.

- Focused on stabilizing the South and protecting African American rights.
- Signed the Civil Rights Act of 1875, aiming to ensure equal treatment in public accommodations.
- Oversaw the passage of the 15th Amendment, securing voting rights for African American men.
- Faced economic turmoil, including the Panic of 1873, which strained his administration.
- Appointed reformers to combat corruption but struggled with scandals like the Whiskey Ring.
- Promoted peace with Native American tribes, though policies often failed to prevent conflict.
- Strengthened foreign policy, resolving disputes with Britain over Civil War claims.

Accomplishments

- Led the Union to victory in the Civil War, preserving the United States.
- Secured ratification of the 15th Amendment, guaranteeing voting rights regardless of race.
- Established the Department of Justice in 1870 to enforce federal laws.

- Created Yellowstone National Park in 1872, the first national park in the U.S.
- Strengthened Reconstruction efforts, using federal troops to protect African Americans.
- Negotiated the Treaty of Washington (1871), peacefully resolving U.S.-British tensions.

First Lady's Contributions

Julia Dent Grant, as First Lady, brought warmth and stability to the White House.

- Hosted lavish receptions, restoring social vibrancy after the somber Lincoln years.
- Supported charitable causes, including aid for Civil War veterans and orphans.
- Promoted women's education, quietly advocating for expanded opportunities.
- Strengthened the Grant family's public image, countering criticism of her husband.

Positive Traits

Grant's personal strengths shaped his leadership, though they sometimes clashed with political realities.

- Determination: Relentless in pursuing goals, from military campaigns to Reconstruction.
- Loyalty: Trusted aides and friends, fostering strong alliances.

- Humility: Remained unpretentious, earning respect from soldiers and citizens.
- Moral conviction: Committed to justice, particularly for African Americans.

Negative Traits

Grant's weaknesses contributed to challenges during his presidency.

- Naivety: Trusted corrupt associates, leading to scandals that tarnished his administration.
- Indecisiveness: Sometimes hesitated on complex policy issues, delaying reforms.
- Inexperience: Lacked political savvy, struggling to navigate Washington's factions.
- Overreliance on military solutions: Applied martial tactics to civil issues, alienating some allies.

Effects of Traits on Presidency

Grant's traits had a mixed impact on his tenure.

- His determination drove Reconstruction forward, but couldn't fully overcome Southern resistance.
- Loyalty led to appointments of unqualified friends, fueling scandals like the Crédit Mobilier affair.
- Humility endeared him to the public but made him underestimate political enemies.
- Naivety allowed corrupt officials to exploit his administration, damaging his legacy.

Pets

Grant, an animal lover, particularly cherished horses, and his family kept several pets.

- Cincinnati: Grant's favorite warhorse, ridden during key Civil War battles.
- Jeff Davis: A horse captured during the war, named ironically after the Confederate president.
- Dogs: The family kept several, including Faithful, a Newfoundland who guarded the White House.
- Ponies: His children had ponies, which Grant occasionally rode with them for fun.

Religious Persuasion

Grant's faith was private but influential, rooted in his upbringing.

- Raised Methodist, attending church sporadically due to military and political demands.
- Respected religious diversity, supporting the separation of church and state.
- Expressed belief in Christian values, emphasizing duty and morality.
- Julia's stronger Methodist faith influenced family practices, including prayer.

Interesting Anecdotes

Grant's life was filled with colorful stories that reveal his character.

- At West Point, he accidentally enrolled as Ulysses S. Grant due to a clerical error, adopting the name permanently.

- During the Civil War, he smoked up to 20 cigars daily, a habit that later contributed to his throat cancer.

- Once, he rode his horse through Washington, D.C., at such speed that he was fined for reckless driving.

- Wrote his memoirs while dying, racing to finish them to secure his family's financial future.

Age at Death

- Ulysses S. Grant died at age 63 on July 23, 1885.

- Julia Dent Grant died at age 76 on December 14, 1902.

Cause of Death

- Ulysses S. Grant: Throat cancer, likely worsened by heavy cigar smoking.

- Julia Dent Grant: Heart failure, compounded by old age and declining health.

Burial Location

- Ulysses S. Grant: Entombed in Grant's Tomb, New York City, a grand mausoleum dedicated in 1897.

- Julia Dent Grant: Buried alongside her husband in Grant's Tomb, New York City.

Age at Death

- Ulysses S. Grant died at age 63 on July 23, 1885.
- Julia Dent Grant died at age 76 on December 14, 1902.
- **Cause of Death**
- Ulysses S. Grant: Throat cancer, likely worsened by heavy cigar smoking.
- Julia Dent Grant: Heart failure, compounded by old age and declining health.

Burial Location

- Ulysses S. Grant: Entombed in Grant's Tomb, New York City, a grand mausoleum dedicated in 1897.
- Julia Dent Grant: Buried alongside her husband in Grant's Tomb, New York City.

Conclusion

Ulysses S. Grant's life was a testament to resilience, from his rise as a Civil War hero to his turbulent presidency. His commitment to Reconstruction and civil rights marked significant progress, though scandals and political inexperience clouded his legacy. Julia's support as First Lady and his personal grit humanized a leader who faced immense challenges. Grant's story reflects the complexities of a nation healing from division, leaving a legacy as both a military genius and a flawed but dedicated president.

Rutherford B. Hayes: Advocate for Civil Service Reform

Introduction

Rutherford B. Hayes, the 19th President of the United States, served from 1877 to 1881. His presidency is most remembered for his efforts to reform the civil service and his role in ending Reconstruction in the South. A lawyer by training and a Republican with strong moral convictions, Hayes navigated a turbulent political landscape marked by a disputed election and lingering post-Civil War tensions. His commitment to integrity and reform shaped his legacy, though his decisions, particularly on Reconstruction, remain controversial.

Early Life

Rutherford Birchard Hayes was born on October 4, 1822, in Delaware, Ohio. His early life was shaped by modest circumstances and a strong emphasis on education.

- His father, Rutherford Hayes Jr., died before his birth, leaving his mother, Sophia Birchard Hayes, to raise him and his siblings.
- Hayes attended local schools and showed an early aptitude for learning, particularly in literature and history.
- He studied at Kenyon College, graduating as valedictorian in 1842, and later attended Harvard Law School, earning his law degree in 1845.
- His early career as a lawyer in Cincinnati established him as a respected figure in Ohio's legal community.

Family

Hayes came from a close-knit family, and his mother's resilience profoundly influenced him.

- Sophia Birchard Hayes, a widow, raised Rutherford and his sister Fanny with a focus on education and moral values.
- His uncle, Sardis Birchard, a successful businessman, provided financial support and served as a father figure.
- Hayes maintained strong ties with his extended family, often relying on their support during his political career.

Children

Rutherford B. Hayes and his wife, Lucy Webb Hayes, had eight children, five of whom survived to adulthood.

- Birchard Austin Hayes (1853–1926), the eldest, became a lawyer and businessman.

- Webb Cook Hayes (1856–1934) served in the military and later worked in business.
- Rutherford Platt Hayes (1858–1927) pursued a career in library science.
- Joseph Thompson Hayes (1861–1863) died in infancy.
- George Crook Hayes (1864–1866) also died young.
- Fanny Hayes (1867–1950) was active in social causes.
- Scott Russell Hayes (1871–1923) worked in banking.
- Manning Force Hayes (1873–1874) died in infancy.
- The Hayes children were raised with a strong sense of duty, reflecting their parents' values.

Rise to Power

Hayes's ascent to the presidency was marked by his legal career, military service, and political roles in Ohio.

- After establishing a law practice in Cincinnati, he gained prominence defending fugitive slaves, showcasing his antislavery stance.
- During the Civil War, he served as a Union officer, rising to the rank of brevet major general and earning respect for his bravery.
- Post-war, Hayes served as a U.S. congressman (1865–1867) and then as Ohio's governor (1868–1872, 1876–1877), where he focused on fiscal responsibility and civil rights.

- His nomination as the Republican candidate in 1876 came after a contentious election, resolved by the Compromise of 1877, which secured his presidency.

Influences

Hayes was shaped by a blend of personal, intellectual, and political influences.

- His mother's Presbyterian values instilled a strong moral compass.
- His education at Kenyon and Harvard exposed him to classical liberalism and legal principles.
- Antislavery advocates, such as Salmon P. Chase, inspired his commitment to civil rights.
- The political climate of post-Civil War America, with its debates over Reconstruction, influenced his focus on national reconciliation and reform.

Party Affiliation

Hayes was a lifelong Republican, aligning with the party's principles during its formative years.

- He joined the Republican Party in the 1850s, drawn to its antislavery platform.
- As a Republican, he supported Union preservation, civil rights for freedmen, and economic development.
- His moderate stance within the party helped him navigate factional divides between Stalwarts and reform-minded Republicans.

Presidency

Hayes's presidency (1877–1881) was defined by efforts to heal a divided nation and reform government practices.

- The Compromise of 1877 ended Reconstruction by withdrawing federal troops from the South, a decision that secured his presidency but allowed Southern states to curtail African American rights.

- He prioritized civil service reform, issuing an executive order to curb political patronage and clashing with party bosses like Roscoe Conkling.

- Hayes advocated for economic stability, supporting the resumption of the gold standard in 1879.

- His administration faced labor unrest, notably the Great Railroad Strike of 1877, which he addressed by deploying federal troops to restore order.

- He promoted education and land grants for Native Americans, though his policies often fell short of meaningful reform.

Accomplishments

- Implemented civil service reforms, including an executive order to promote merit-based appointments.

- Oversaw the resumption of the gold standard, stabilizing the economy.

- Ended Reconstruction, aiming for national reconciliation, though this led to the erosion of African American rights.

- Vetoed the Bland-Allison Act to limit silver coinage, though Congress overrode his veto.
- Promoted federal funding for internal improvements, such as harbor and river projects.

First Lady's Contributions

Lucy Webb Hayes, known as "Lemonade Lucy" for her temperance stance, was an influential first lady.

- Advocated for temperance, banning alcohol in the White House, which shaped its social atmosphere.
- Supported education and veterans' welfare, visiting schools and hospitals.
- Hosted inclusive White House events, fostering a welcoming environment for diverse guests.
- Promoted women's education, reflecting her own status as a college graduate.
- Influenced her husband's policies through her moral and humanitarian perspective.

Positive Traits

Hayes's positive traits significantly shaped his presidency.

- His integrity and commitment to reform earned him respect, even from political opponents.
- A strong work ethic and attention to detail allowed him to tackle complex issues like civil service reform.

- His compassion, rooted in his antislavery background, informed his efforts to promote education and Native American welfare.

- These traits helped him maintain public trust and push for long-term governance improvements, despite resistance.

Negative Traits

Hayes's flaws also impacted his presidency.

- His cautious and conciliatory nature led to compromises, such as ending Reconstruction, that alienated African Americans and reformers.

- A lack of political assertiveness sometimes weakened his ability to counter powerful party factions.

- His idealism occasionally outpaced practical outcomes, limiting the effectiveness of his policies.

- These traits contributed to a presidency that, while principled, struggled to achieve lasting change in some areas.

Pets

The Hayes family brought several pets to the White House, reflecting their lively household.

- They owned a greyhound named Grim, a mastiff named Juno, and a Newfoundland named Hector.

- Cats, including a Siamese named Siam, were among the first of their breed in the U.S.

- They also kept birds, such as canaries and mockingbirds, adding to the White House's vibrant atmosphere.

Religious Persuasion

Hayes and his wife were devout Christians, with their faith guiding their personal and public lives.

- Raised in the Presbyterian Church, Hayes later aligned with Methodism, Lucy's denomination.
- Their faith emphasized moral duty, temperance, and charity, influencing White House policies like the alcohol ban.
- Hayes attended church regularly and supported religious organizations, though he avoided imposing his beliefs politically.

Interesting Anecdotes

Hayes's life was marked by unique moments that highlighted his character.

- During the Civil War, he was wounded five times but continued leading his troops, earning the nickname "Old Bulletproof."
- He and Lucy were the first presidential couple to host an Easter Egg Roll on the White House lawn in 1878, a tradition that endures.
- Hayes kept a diary throughout his life, offering historians a detailed glimpse into his thoughts and decisions.

- He was the first president to use a telephone in the White House, installed by Alexander Graham Bell in 1879.

Ages at Death

- Rutherford B. Hayes died at age 70 on January 17, 1893.
- Lucy Webb Hayes died at age 57 on June 25, 1889.

Causes of Death

- Hayes suffered a heart attack, likely exacerbated by a cold contracted during a trip to Cleveland.
- Lucy died of a stroke, following a period of declining health.

Burial Locations

- Both Rutherford and Lucy Hayes are buried at Spiegel Grove, their estate in Fremont, Ohio.
- The site, now part of the Rutherford B. Hayes Presidential Library and Museums, includes their tomb and a memorial.

Conclusion

Rutherford B. Hayes's presidency, though often overshadowed by more prominent figures, was a pivotal moment in American history. His commitment to civil service reform laid the groundwork for modern governance, while his decision to end Reconstruction reflected the complex challenges of national unity. Lucy Hayes complemented her husband's efforts with her advocacy for temperance and education, leaving a lasting mark on the role of the first lady. Despite controversies, Hayes's integrity and dedication to public service defined his legacy as a principled leader in a divided era.

James A. Garfield: The Assassinated Reformer

Introduction

James Abram Garfield, the 20th President of the United States, served a brief but impactful term in 1881 before his assassination cut his presidency short. A scholar, soldier, and statesman, Garfield rose from humble beginnings to become a champion of civil rights and political reform. His life reflects a blend of intellectual rigor, moral conviction, and resilience, though his presidency was marred by political infighting and tragedy.

Early Life

James Garfield was born on November 19, 1831, in a log cabin in Orange Township, Ohio. His early years were marked by poverty and hardship, shaping his work ethic and empathy for the disadvantaged.

- Grew up in a frontier environment after his father's death when Garfield was 18 months old.

- Raised by his mother, Eliza Ballou Garfield, who instilled a love of learning.
- Worked as a canal boat driver and farmhand to support his family as a teenager.
- Attended Geauga Academy and later the Western Reserve Eclectic Institute (now Hiram College), excelling in classics, mathematics, and rhetoric.
- Graduated from Williams College in Massachusetts in 1856, where he honed his oratorical skills and intellectual curiosity.

Family

Garfield's family life was anchored by his strong partnership with his wife, Lucretia, and their shared commitment to education and public service.

- Married Lucretia Rudolph on November 11, 1858, after a long courtship.
- Their marriage was initially reserved but grew into a deep, supportive partnership.
- Lucretia was a stabilizing force, managing the household during Garfield's frequent absences for political and military duties.
- The couple maintained close ties with Garfield's mother, Eliza, who lived with them periodically.
- Garfield valued family life, often writing affectionate letters to Lucretia and their children during his travels.

Children

James and Lucretia Garfield had seven children, though two died in infancy. Their surviving children were active in public life, reflecting their parents' values.

- Harry Augustus Garfield (1863–1942): Became president of Williams College and a prominent educator.
- James Rudolph Garfield (1865–1950): Served as U.S. Secretary of the Interior under Theodore Roosevelt.
- Mary "Mollie" Garfield (1867–1947): Married Joseph Stanley-Brown, Garfield's private secretary, and preserved her father's papers.
- Irvin McDowell Garfield (1870–1951): Became a lawyer in Boston.
- Abram Garfield (1872–1958): Worked as an architect in Cleveland.
- Two children, Eliza Arabella (1860–1863) and Edward (1874–1876), died young, deeply affecting the family.

Rise to Power

Garfield's ascent to the presidency was marked by his intellectual prowess, military service, and political acumen, despite his initial reluctance to seek high office.

- Began as a teacher and principal at the Western Reserve Eclectic Institute, gaining a reputation as a gifted educator.

- Elected to the Ohio State Senate in 1859 as a Republican, advocating for anti-slavery policies.
- Served as a Union general during the Civil War, distinguishing himself at battles like Shiloh and Chickamauga.
- Elected to the U.S. House of Representatives in 1863, serving nine terms and becoming a leading voice on fiscal policy and Reconstruction.
- Nominated as a compromise candidate for president at the 1880 Republican National Convention, defeating Ulysses S. Grant's faction.

Influences

Garfield's worldview was shaped by a blend of intellectual, moral, and political influences that guided his reformist agenda.

- Inspired by the abolitionist movement and figures like Frederick Douglass, he shaped his commitment to civil rights.
- Influenced by his classical education at Williams College, particularly the works of Cicero and Aristotle, which informed his oratory.
- Mentored by political figures like Salmon P. Chase, who reinforced his belief in fiscal responsibility and anti-corruption measures.
- Drew on his military experience to advocate for national unity and Reconstruction policies.

- Shaped by his mother's resilience and faith, which instilled a sense of duty and humility.

Party Affiliation

Garfield was a steadfast member of the Republican Party, aligning with its progressive and reformist wings during a turbulent era.

- Joined the Republican Party in the 1850s, drawn to its anti-slavery platform.
- Supported the "Radical Republicans" during Reconstruction, advocating for African American voting rights.
- Navigated intra-party conflicts between Stalwarts (favoring patronage) and Half-Breeds (favoring reform), leaning toward the latter.
- His 1880 presidential campaign unified a fractured Republican Party, balancing competing factions.
- Championed civil service reform to curb the spoils system, a stance that alienated some party loyalists.

Presidency

Garfield's presidency, from March 4, 1881, to September 19, 1881, was cut short by assassination but showed promise for significant reform.

- Focused on strengthening the Union and advancing civil rights for freedmen.
- Confronted the spoils system, pushing for merit-based appointments in government.

- Strengthened U.S. naval power and advocated for modernizing the military.
- Appointed African Americans to prominent federal positions, including Frederick Douglass as Recorder of Deeds.
- Faced resistance from Stalwart Republicans, particularly Senator Roscoe Conkling, over patronage disputes.
- Shot by Charles J. Guiteau, a disgruntled office seeker, on July 2, 1881, and lingered for 80 days before succumbing to infection.

Accomplishments

- Initiated civil service reform, laying the groundwork for the Pendleton Act of 1883.
- Appointed diverse and qualified individuals to federal posts, promoting inclusivity.
- Strengthened U.S. foreign policy by supporting Pan-American cooperation.
- Advanced fiscal policies to stabilize the post-Civil War economy.
- Championed education and literacy programs for African Americans during Reconstruction.

First Lady's Contributions

- Hosted intellectual salons at the White House, fostering cultural and political discourse.

- Supported educational initiatives, particularly for women and freedmen.

- Preserved Garfield's legacy by curating his papers and correspondence after his death.

- Advocated for historical preservation, contributing to the establishment of the James A. Garfield National Historic Site.

- Provided emotional support to Garfield during his presidency, managing family affairs under public scrutiny.

Positive Traits

Garfield's personal strengths shaped his leadership but were tempered by the brevity of his term.

- Intellectual curiosity: A voracious reader and scholar, he brought depth to policy debates.

- Eloquence: His oratorical skills rallied support for reform and unity.

- Empathy: His humble roots fostered a genuine concern for the disadvantaged.

- Courage: His Civil War service and willingness to confront party bosses demonstrated resolve.

- These traits inspired trust and laid the foundation for his reform agenda, though his assassination limited their impact.

Negative Traits

Garfield's flaws, while minor, created challenges during his presidency.

- Indecisiveness: His tendency to deliberate extensively sometimes delays action.
- Overly conciliatory: Efforts to appease party factions led to tensions with Stalwarts.
- Inexperience in executive roles: His brief presidency exposed a learning curve in managing federal bureaucracy.
- These traits exacerbated conflicts with political opponents, contributing to the patronage disputes that indirectly led to his assassination.

Pets

The Garfield family kept pets, reflecting their domestic warmth and love for animals.

- Owned a dog named Veto, a Newfoundland, known for its loyalty and gentle temperament.
- Veto often accompanied Garfield on walks and was a favorite of the children.
- The family also kept horses for transportation and recreation at their Mentor, Ohio, home.
- Pets provided comfort during Garfield's demanding political career and Lucretia's White House tenure.

Religious Persuasion

Garfield's faith was a cornerstone of his moral framework, influencing his public and private life.

- Raised in the Disciples of Christ (Campbellite) church, emphasizing simplicity and personal faith.
- Served as a lay preacher in his youth, delivering sermons at local congregations.
- Believed in the separation of church and state but drew on Christian ethics to guide his policies.
- Lucretia shared his faith, and they raised their children in the Disciples of Christ tradition.
- His religious convictions reinforced his commitment to justice and equality, particularly for African Americans.

Interesting Anecdotes

Garfield's life was filled with unique moments that highlighted his character and intellect.

- As a young canal worker, he nearly drowned but taught himself to swim, showcasing his resilience.
- Could write Latin with one hand and Greek with the other simultaneously, a skill he demonstrated to impress colleagues.
- During the Civil War, he debated theology with a Confederate prisoner, earning mutual respect.
- Refused a lucrative legal case as a congressman to avoid conflicts of interest, reflecting his integrity.

- His assassination sparked national mourning and accelerated civil service reform, a bittersweet legacy.

Ages at Death

- James A. Garfield died at age 49 on September 19, 1881.
- Lucretia Garfield died at age 85 on March 14, 1918.
- Garfield's early death was a national tragedy, while Lucretia's long life allowed her to preserve his legacy.

Causes of Death

- James A. Garfield: Died from infections caused by poor medical treatment following his assassination by gunshot on July 2, 1881.
- Lucretia Garfield: Died of natural causes, likely related to old age, in Pasadena, California.
- Garfield's prolonged suffering highlighted the era's limited medical knowledge, while Lucretia's longevity reflected her resilience.

Burial Locations

- James A. Garfield: Buried at Lake View Cemetery in Cleveland, Ohio, in a grand monument dedicated in 1890.
- Lucretia Garfield: Also buried at Lake View Cemetery, alongside her husband, in the Garfield Memorial.
- The monument, a 180-foot Romanesque structure, remains a pilgrimage site for history enthusiasts.

Conclusion

James A. Garfield's life was a testament to the power of intellect, perseverance, and moral conviction. From a log cabin to the White House, he overcame adversity to champion civil rights and reform. His brief presidency, though tragically cut short, planted seeds for civil service reform and national unity. Lucretia Garfield's contributions as a supportive partner and cultural advocate enriched their legacy. Despite his flaws, Garfield's empathy and vision left an indelible mark on American history, remembered most as the assassinated reformer whose potential was never fully realized. His story reminds us of the fragility of progress and the enduring impact of principled leadership.

Chester A. Arthur: The Unexpected Reformer

Description

Chester Alan Arthur, the 21st President of the United States, was a polished, dignified man who rose from humble beginnings to become a key figure in American politics. Known for his unexpected commitment to civil service reform, Arthur defied expectations of a corrupt political operative to leave a lasting legacy of integrity in governance.

Introduction

Chester A. Arthur served as president from September 19, 1881, to March 4, 1885, assuming office after the assassination of James A. Garfield. Initially viewed as a political insider tied to the spoils system, Arthur surprised critics by championing reforms that transformed the federal government. His presidency, though brief, marked a turning point in addressing corruption and inefficiency in public administration.

Early Life

Born on October 5, 1829, in Fairfield, Vermont, Chester A. Arthur grew up in a modest, religious household. His early years shaped his character and ambitions, setting the stage for his legal and political career.

- Son of a Baptist preacher, William Arthur, and Malvina Stone Arthur.
- Educated at local schools and later at Union College in Schenectady, New York, graduating in 1848.
- Taught school briefly before pursuing law, admitted to the New York bar in 1854.
- Developed an early interest in abolitionism, influenced by his father's anti-slavery sermons.

Family

Arthur's family life was centered around his wife, Ellen, and their children, though tragedy marked their later years.

- Married Ellen Lewis Herndon in 1859, a Virginia-born woman from a prominent family.
- Close relationship with his parents and siblings, maintaining ties despite his demanding career.
- Ellen's Southern roots created occasional social tensions during the Civil War era.
- Arthur was a devoted family man, though his political duties often kept him away from home.

Children

Chester and Ellen Arthur had three children, one of whom died young, shaping Arthur's private life with both joy and sorrow.

- William Lewis Arthur (1860–1863): Died at age two from a brain ailment, deeply affecting the family.
- Chester Alan Arthur II (1864–1937): Known as "Allan," pursued a leisurely life, avoiding politics.
- Ellen Herndon Arthur (1871–1915): Known as "Nell," lived a private life, later marrying.

Rise to Power

Arthur's ascent in politics was tied to his legal acumen and connections within the Republican Party, particularly in New York's political machine.

- Began as a lawyer in New York City, handling notable civil rights cases, including defending fugitive slaves.
- Joined the Republican Party in the 1850s, aligning with powerful New York senator Roscoe Conkling.
- Appointed Collector of the Port of New York in 1871, a lucrative post where he managed patronage.
- Selected as Garfield's vice-presidential running mate in 1880 to balance the ticket, assuming the presidency after Garfield's death.

Influences

Arthur's worldview was shaped by a blend of personal, professional, and political experiences.

- His father's abolitionist preaching instilled a sense of justice, evident in his early legal work.
- Roscoe Conkling's mentorship provided political savvy but tied Arthur to the spoils system.
- Exposure to New York's diverse society broadened his perspective on governance.
- Garfield's assassination and public outcry for reform pushed Arthur toward unexpected integrity.

Party Affiliation

- Republican Party: Arthur was a loyal Republican, initially aligned with the Stalwart faction, which favored patronage.
- Shifted toward reform-minded policies during his presidency, distancing himself from Conkling's machine politics.
- His independent streak frustrated party loyalists but earned him respect for prioritizing national interest.

Presidency

Arthur's presidency, though initially met with skepticism, is remembered for significant reforms and steady leadership during a time of political turmoil.

- Assumed office after Garfield's assassination, facing distrust due to his patronage background.
- Championed civil service reform, defying expectations of a machine politician.

- Oversaw modernization of the U.S. Navy, strengthening national defense.
- Vetoed the Chinese Exclusion Act in 1882 for being too harsh, but signed a revised version, reflecting complex immigration debates.
- Managed economic growth and federal surplus, though struggled with tariff reform.

Accomplishments

- Signed the Pendleton Civil Service Reform Act (1883), establishing merit-based federal hiring.
- Modernized the U.S. Navy, commissioning steel warships to bolster maritime strength.
- Improved federal infrastructure, including post offices and public buildings.
- Promoted fair administration of customs and tax collection, reducing corruption.
- Vetoed excessive pork-barrel spending, showing fiscal restraint.

First Lady's Contributions

Ellen Arthur died before Chester became president, leaving no official First Lady during his term. His sister, Mary Arthur McElroy, served as White House hostess.

- Mary McElroy hosted social events, maintaining White House traditions with grace.
- Organized receptions to foster diplomatic and political connections.

- Supported Arthur emotionally, stepping in during his grief over Ellen's death.
- Promoted cultural events, enhancing the White House's social prestige.

Positive Traits

- Polished and diplomatic, Arthur excelled at building relationships across political divides.
- Adaptable, shifting from a patronage loyalist to a reform advocate when faced with public demand.
- Dignified demeanor earned respect, projecting stability during a turbulent period.
- Legal acumen informed his governance, ensuring thoughtful policy decisions.
- These traits enabled the passage of the Pendleton Act and restored public trust in his leadership.

Negative Traits

- Initially tied to the corrupt spoils system, which fueled public skepticism.
- Reserved nature sometimes appeared aloof, limiting his ability to inspire.
- Reluctance to confront party bosses early in his career delayed his reformist turn.
- Health issues, including Bright's disease, reduced his energy and focus later in his term.

- These traits initially hindered his credibility but were largely overcome by his reform efforts.

Pets

- Arthur was not known to keep pets in the White House, focusing instead on his duties and social obligations.
- His household likely included no notable animals, reflecting his urban lifestyle and lack of documented interest in pets.

Religious Persuasion

- Raised in a devout Baptist family, influenced by his father's preaching.
- Later identified as Episcopalian, attending services in New York and Washington, D.C.
- Religion played a private role in his life, with no significant public displays during his presidency.
- His faith likely reinforced his sense of duty and moral responsibility in office.

Interesting Anecdotes

- Arthur was known for his impeccable fashion, earning the nickname "Elegant Arthur" for his stylish wardrobe.
- He reportedly stayed up late hosting lavish White House dinners, reflecting his love for socializing.
- As a lawyer, he won a case in 1855 that desegregated New York City streetcars, a lesser-known civil rights achievement.

- Despite his reformist presidency, he burned many personal papers before his death, leaving historians with limited insights into his private thoughts.

Age at Death

- Chester A. Arthur: Died at age 57 on November 18, 1886.
- Ellen Arthur: Died at age 42 on January 12, 1880.

Cause of Death

- Chester A. Arthur: Died of Bright's disease, a kidney ailment, exacerbated by the stresses of his presidency.
- Ellen Arthur: Died of pneumonia, contracted after a cold worsened during a winter concert.

Burial Location

- Chester A. Arthur: Buried at Albany Rural Cemetery, Menands, New York, in a simple family plot.
- Ellen Arthur: Also buried at Albany Rural Cemetery, alongside her husband and family.

Conclusion

Chester A. Arthur's presidency, though born of tragedy, marked a pivotal shift toward modern governance. His unexpected embrace of civil service reform, particularly the Pendleton Act, curbed the excesses of the spoils system and laid the foundation for a merit-based bureaucracy. Despite personal health struggles and the loss of his wife, Arthur's dignity, adaptability, and commitment to reform earned him a respected place in history. His legacy as the

"Unexpected Reformer" endures, proving that even unlikely leaders can rise to meet the demands of their time.

Grover Cleveland: The First President to Serve Two Non-Consecutive Terms

Introduction

Grover Cleveland, the 22nd and 24th President of the United States, holds a unique place in American history as the only president to serve two non-consecutive terms (1885–1889 and 1893–1897). Known for his steadfast commitment to fiscal conservatism and political integrity, Cleveland navigated a rapidly industrializing nation through economic crises and social changes. His presidency was marked by a dedication to limited government, resistance to corruption, and a focus on restoring public trust in governance. Despite his principled leadership, Cleveland faced significant challenges, including economic depressions and labor unrest, which shaped his legacy as a resolute but sometimes polarizing figure.

Description

Stephen Grover Cleveland, commonly known as Grover Cleveland, was a robust, plain-spoken man with a reputation for honesty and independence. Born on March 18, 1837, in Caldwell, New Jersey, he rose from humble beginnings to become a dominant figure in American politics. His imposing physical presence—standing at 5 feet 11 inches and weighing over 250 pounds—matched his forceful personality. Cleveland's no-nonsense demeanor and commitment to principle earned him the nickname "Guardian President," though his stubbornness sometimes alienated allies and opponents alike.

Early Life Cleveland grew up in a modest, religious household, the fifth of nine children. His early years were shaped by hard work and a strong sense of duty.

- Born in Caldwell, New Jersey, to a Presbyterian minister father and a devout mother.
- Moved to Fayetteville, New York, at age four, where he attended local schools.
- Left school at 16 after his father's death to support his family, working as a clerk and teacher.
- Moved to Buffalo, New York, in 1855, where he studied law and was admitted to the bar in 1859.
- Developed a reputation as a diligent lawyer, known for his work ethic and attention to detail.

Family

Cleveland's family life was unconventional for a president, as he married during his first term and started a family later in life.

- Parents: Richard Falley Cleveland, a Presbyterian minister, and Ann Neal Cleveland.
- Siblings: Eight, including his sister Rose, who briefly served as White House hostess.
- Married Frances Folsom in 1886, the daughter of his late law partner, in a White House ceremony—the only presidential wedding held there.
- Frances, 27 years his junior, was 21 at the time of their marriage, sparking public fascination.

Children

Cleveland and Frances had five children, four of whom survived to adulthood. Their family life was closely scrutinized due to Cleveland's prominence and the age difference in their marriage.

- Ruth Cleveland (1891–1904) died young of diphtheria; inspired the "Baby Ruth" candy bar name.
- Esther Cleveland (1893–1980) was the only presidential child born in the White House.
- Marion Cleveland (1895–1977) later active in education and philanthropy.
- Richard Folsom Cleveland (1897–1974) became a lawyer.

- Francis Grover Cleveland (1903–1995) pursued a career in theater.

Rise to Power

Cleveland's ascent to the presidency was swift, driven by his reputation for integrity and reform in local and state politics.

- Began as a lawyer in Buffalo, gaining prominence for his legal acumen.
- Elected sheriff of Erie County in 1870, earning respect for his fairness and efficiency.
- Became mayor of Buffalo in 1882, where he fought corruption and earned the nickname "Veto Mayor."
- Elected governor of New York in 1882, continuing his reform agenda by challenging Tammany Hall's political machine.
- Nominated as the Democratic candidate for president in 1884, capitalizing on his anti-corruption stance.

Influences

Cleveland's political philosophy was shaped by his upbringing and professional experiences.

- His father's Presbyterian values instilled a strong sense of duty and morality.
- Exposure to Buffalo's political corruption as a young lawyer fueled his commitment to reform.
- Influenced by Democratic principles of limited government and fiscal responsibility.

- Admired Thomas Jefferson's ideals of individual liberty and restrained federal power.
- Mentored by Oscar Folsom, his law partner, whose death led Cleveland to become guardian to Frances Folsom.

Party Affiliation

Cleveland was a steadfast Democrat, aligning with the party's conservative wing during an era of intense political division.

- Supported the Democratic Party's emphasis on limited government and low tariffs.
- Opposed the Republican Party's protectionist policies and expansive federal spending.
- Criticized by some Democrats for his independence, as he often prioritized principle over party loyalty.
- His anti-corruption stance appealed to reform-minded voters across party lines, including "Mugwumps" who supported him in 1884.

Presidency

Cleveland's two terms were defined by economic challenges, reform efforts, and a commitment to principled governance, though his policies often sparked controversy.

- First term (1885–1889): Focused on reducing government waste, vetoing excessive pension bills, and promoting civil service reform.

- Fought against high tariffs, advocating for free trade to lower consumer prices.
- Lost re-election in 1888 to Benjamin Harrison, despite winning the popular vote, due to electoral college dynamics.
- Second term (1893–1897): Faced the Panic of 1893, a severe economic depression, and supported the gold standard, alienating pro-silver Democrats.
- Sent federal troops to break the Pullman Strike in 1894, a controversial decision that damaged his popularity among labor groups.
- Strengthened the Monroe Doctrine by intervening in the Venezuela-Britain boundary dispute, asserting U.S. influence in the Western Hemisphere.

Accomplishments

- Vetoed over 400 bills in his first term, a record at the time, to curb wasteful spending.
- Signed the Interstate Commerce Act of 1887, the first federal regulation of railroads.
- Promoted civil service reform through the Pendleton Act's expansion, reducing patronage in government jobs.
- Repealed the Tenure of Office Act, strengthening presidential authority.
- Established the Department of Agriculture as a Cabinet-level agency in 1889.

- Resolved the Venezuela Crisis of 1895, reinforcing U.S. foreign policy dominance in the Americas.

First Lady's Contributions

- Frances Cleveland modernized the role of First Lady, becoming a public icon for her youth and charm.
- Hosted public receptions to make the White House more accessible to ordinary citizens.
- Advocated for women's education, supporting institutions like Wells College, her alma mater.
- Influenced fashion trends, with her style widely emulated by American women.
- Supported charitable causes, including aid for impoverished families in Washington, D.C.
- Remained a beloved figure post-presidency, raising funds for education and veterans' causes.

Positive Traits

Cleveland's strengths shaped his leadership but sometimes limited his political agility.

- Uncompromising integrity: Refused to bend to political pressure, earning public trust.
- Work ethic: Known for long hours and meticulous attention to policy details.
- Courage: Took bold stands, such as opposing inflationary silver policies, despite party opposition.

- These traits bolstered his reputation as a principled leader but made compromise difficult, alienating allies during economic crises.

Negative Traits

Cleveland's flaws often complicated his presidency, particularly in times of crisis.

- Stubbornness: His refusal to adapt policies, like his rigid support for the gold standard, deepened economic woes.

- Aloofness: Struggled to connect emotionally with the public, appearing distant during labor unrest.

- Limited political vision: Focused on fiscal restraint over broader social reforms, missing opportunities to address industrialization's challenges.

- These traits fueled criticism, particularly during the Panic of 1893, when his unyielding policies exacerbated public discontent.

Pets

Cleveland and his family kept a variety of pets, reflecting their domestic life.

- Owned several dogs, including a Japanese poodle named Hector.

- Kept canaries and other birds, gifts from admirers, in the White House.

- Frances maintained a small collection of pet fish, adding a personal touch to the executive mansion.

Religious Persuasion

Cleveland's faith influenced his moral outlook but was not overtly public.

- Raised in a devout Presbyterian household, attending church regularly as a young man.
- Maintained personal religious beliefs but avoided emphasizing religion in his political life.
- Frances was also Presbyterian, and the couple raised their children in the faith, though they were not overtly evangelical.

Interesting Anecdotes

- During the 1884 campaign, Cleveland faced a scandal over fathering an illegitimate child; he admitted responsibility, defusing the issue with honesty.
- As sheriff, he personally performed two hangings, a duty he took seriously to uphold justice.
- His White House wedding to Frances Folsom captivated the nation, with crowds gathering outside to celebrate.
- Cleveland once answered the White House telephone himself, surprising callers who expected a staffer.
- He secretly underwent surgery for mouth cancer in 1893 aboard a yacht to avoid public panic, with the procedure kept hidden for decades.

Ages at Death

- Grover Cleveland died at age 71 on June 24, 1908.
- Frances Cleveland died at age 83 on October 29, 1947.

Causes of Death

- Grover Cleveland succumbed to heart failure, complicated by his history of obesity and gout.
- Frances Cleveland died of natural causes related to old age.

Burial Locations

- Grover Cleveland was buried in Princeton Cemetery, Princeton, New Jersey, near his retirement home.
- Frances Cleveland was also buried in Princeton Cemetery, alongside her husband, after her death in 1947.

Conclusion

Grover Cleveland's legacy as the only president to serve two non-consecutive terms reflects his enduring commitment to integrity and fiscal conservatism. His leadership during turbulent economic times showcased both his strengths—unwavering principle and dedication to reform—and his weaknesses, including stubbornness and a limited vision for social change. Frances Cleveland complemented her husband's presidency with her grace and public engagement, redefining the First Lady's role. Together, they navigated personal and political challenges, leaving a lasting mark on American history through their contributions to governance, diplomacy, and public life.

Cleveland's steadfast adherence to his principles, for better or worse, remains a defining feature of his unique place in the presidency.

Benjamin Harrison: The Centennial President

Description

Benjamin Harrison, the 23rd President of the United States, served from 1889 to 1893. Known as the "Centennial President" for his term coinciding with the 100th anniversary of George Washington's inauguration, Harrison was a principled lawyer, Civil War veteran, and Republican statesman. His presidency focused on economic reforms, civil rights, and foreign policy, though his reserved demeanor often overshadowed his achievements.

Introduction

Benjamin Harrison's legacy is rooted in his commitment to Republican ideals and his efforts to modernize the American economy during a transformative era. A skilled orator and meticulous administrator, he navigated a rapidly industrializing nation while advocating for protective tariffs and civil rights for African Americans. Despite his accomplishments, his single term was marked by political challenges and economic turbulence, leaving a mixed legacy.

Early Life

Born on August 20, 1833, in North Bend, Ohio, Benjamin Harrison grew up in a prominent political family. His grandfather, William Henry Harrison, was the ninth U.S. President, and his great-grandfather, Benjamin Harrison V, signed the Declaration of Independence. Raised on a farm, Harrison developed a strong work ethic and intellectual curiosity. He attended Farmer's College and later Miami University in Ohio, graduating in 1852 with distinction. His early career as a lawyer in Indianapolis established his reputation for integrity and diligence.

Family

Harrison's family ties shaped his values and political career. His father, John Scott Harrison, was a U.S. Congressman, and his mother, Elizabeth Irwin Harrison, instilled Presbyterian values. Harrison married twice, first to Caroline Scott in 1853, who was his partner during his rise to prominence, and later to Mary Dimmick in 1896 after Caroline's death. His family's political legacy provided both opportunity and pressure to uphold a tradition of public service.

Children

Benjamin and Caroline Harrison had three children, two of whom survived to adulthood.

- Russell Benjamin Harrison (1854–1936): Became a businessman and served in minor political roles.
- Mary Scott Harrison (1858–1930): Known as Mamie, she married James McKee and lived a private life.
- An unnamed daughter (1861): Died at birth, a personal tragedy for the Harrisons.
  The children maintained low public profiles, with Russell occasionally involved in his father's political activities.

Rise to Power

Harrison's ascent to the presidency was marked by legal and political success. After establishing a law practice in Indianapolis, he gained prominence as a courtroom advocate. He served as a Union Army officer during the Civil War, rising to the rank of brigadier general. Elected to the U.S. Senate in 1881, he championed protective tariffs and veterans' pensions. His 1888 presidential campaign, leveraging his family name and Republican support, narrowly defeated incumbent Grover Cleveland in the electoral college, despite losing the popular vote.

Influences

Harrison's Presbyterian faith and family legacy profoundly shaped his worldview. His legal training emphasized precision and fairness, while his Civil War service instilled a sense of duty. Influenced by Republican leaders like James G. Blaine, he embraced economic nationalism and protective tariffs. His commitment to civil rights for African Americans reflected his moral convictions, though political realities limited his success in this area.

Party Affiliation

Harrison was a lifelong Republican, aligned with the party's post-Civil War focus on industrial growth, protective tariffs, and federal authority. He supported the party's "bloody shirt" strategy, emphasizing Union loyalty, and advocated for economic policies favoring American businesses. His loyalty to Republican principles sometimes alienated moderates, contributing to his 1892 election loss.

Presidency

Harrison's presidency (1889–1893) addressed economic, foreign, and social issues during a time of industrial growth and political division. His administration faced a divided Congress and economic challenges, including the 1890 financial panic. Key legislation and policies defined his term, though his reserved nature limited his public appeal.

Accomplishments

- Signed the McKinley Tariff Act (1890), raising tariffs to protect U.S. industries.

- Approved the Sherman Antitrust Act (1890), the first federal law to regulate monopolies.
- Supported the Land Revision Act (1891), creating national forest reserves.
- Oversaw the admission of six new states (North Dakota, South Dakota, Montana, Washington, Idaho, Wyoming).
- Strengthened the U.S. Navy, modernizing it into a global force.
- Advocated for federal voting rights protections for African Americans, though bills failed in Congress.
- Hosted the first Pan-American Conference (1889), fostering hemispheric cooperation.

First Lady's Contributions

Caroline Scott Harrison, First Lady from 1889 to 1892, was an active figure despite health challenges.

- Founded the Daughters of the American Revolution (DAR), serving as its first President General.
- Oversaw White House renovations, introducing electricity and modern plumbing.
- Promoted women's education and cultural events, hosting art and music gatherings.
- Supported charitable causes, including orphanages and veterans' families.

Positive Traits and Effects

Harrison's intelligence, integrity, and work ethic earned respect among colleagues. His legal acumen ensured careful policy deliberation, particularly in antitrust and tariff legislation. His commitment to civil rights, though unsuccessful, reflected moral courage. These traits strengthened his administration's legislative output but were often overshadowed by his inability to connect with the public, limiting his political influence.

Negative Traits and Effects

Harrison's aloof demeanor and lack of charisma alienated voters and party allies. His rigid adherence to principle sometimes hindered compromise, exacerbating tensions with Congress. His focus on tariffs contributed to economic discontent, fueling Democratic gains in the 1890 midterms and his 1892 defeat. These traits made his presidency appear distant and unresponsive to public needs.

Pets

The Harrisons kept several pets, reflecting their family-oriented White House.

- Whiskers, a goat, pulled a cart for Harrison's grandchildren.

- Several dogs, including Dash, a collie, roamed the White House grounds.

- A parrot, known for mimicking voices, entertained guests.
  These pets humanized the reserved Harrison, delighting visitors and family.

Religious Persuasion

Harrison was a devout Presbyterian, attending services regularly at the First Presbyterian Church in Washington, D.C. His faith guided his moral decisions, particularly his support for civil rights and temperance. He viewed public service as a divine calling, though he avoided overt religious rhetoric in politics to maintain broad appeal.

Interesting Anecdotes

During his 1888 campaign, Harrison's "front porch" speeches in Indianapolis drew thousands, showcasing his oratorical skill. He once humorously settled a White House staff dispute by citing legal precedent, reflecting his lawyerly instincts. Caroline's White House Christmas decorations, featuring a tree with electric lights, were among the first of their kind, captivating the public.

Ages at Death, Causes of Death, and Burial Locations

- **Benjamin Harrison**: Died at age 67 on March 13, 1901, from pneumonia in Indianapolis, Indiana. Buried at Crown Hill Cemetery, Indianapolis.

- **Caroline Harrison**: Died at age 60 on October 25, 1892, from tuberculosis in the White House, Washington, D.C. Buried at Crown Hill Cemetery, Indianapolis.

- **Mary Dimmick Harrison** (second wife): Died at age 89 on January 5, 1948, from natural causes in New York City. Buried at Crown Hill Cemetery, Indianapolis.

Conclusion

Benjamin Harrison's presidency, though overshadowed by economic challenges and his reserved nature, left a lasting impact through landmark legislation like the Sherman Antitrust Act and naval modernization. His commitment to civil rights, though unrealized, showed foresight. Caroline Harrison's contributions, from founding the DAR to modernizing the White House, complemented his legacy. As the Centennial President, Harrison bridged an era of tradition and progress, leaving a legacy of principled leadership despite political setbacks.

Grover Cleveland: (2nd Term) The First President to Serve Non-Consecutive Terms

Introduction

Grover Cleveland, the 22nd and 24th President of the United States, holds a unique place in American history as the only president to serve two non-consecutive terms (1885–1889 and 1893–1897). Known for his steadfast commitment to fiscal conservatism and political integrity, Cleveland navigated a rapidly industrializing nation through economic crises and political challenges. His presidency is often remembered for his dedication to limited government and his efforts to maintain the gold standard, though his policies and personal traits sparked both admiration and controversy.

Early Life

Stephen Grover Cleveland was born on March 18, 1837, in Caldwell, New Jersey, to a modest family with deep New England roots. His early years shaped his disciplined and principled character, setting the stage for his political career.

- Grew up in a Presbyterian minister's household, the fifth of nine children.

- Moved to Fayetteville, New York, at age four, where he attended local schools.

- Faced financial hardship after his father's death in 1853, forcing him to leave school at 16 to support his family.

- Worked as a teacher in a school for the blind in New York and later as a clerk in Buffalo, studying law in his spare time.

Family

Cleveland's family life was marked by his late marriage and the public fascination with his young wife, Frances Folsom. His personal life, while generally private, became a focal point during his presidency due to his unconventional path to marriage.

- Married Frances Folsom in 1886, the daughter of his late law partner, Oscar Folsom.

- Cleveland was 49 and Frances was 21 at the time of their wedding, the first presidential wedding held in the White House.
- Remained close to his siblings, particularly his sister Rose, who served as acting First Lady before his marriage.

Children

Cleveland and Frances had five children, three daughters and two sons, who were raised in the public eye due to their father's prominence. Their family life was closely scrutinized, adding to Cleveland's public persona.

- Ruth Cleveland (1891–1904), nicknamed "Baby Ruth," inspired the candy bar name (though not officially).
- Esther Cleveland (1893–1980) was the only presidential child born in the White House.
- Marion Cleveland (1895–1977) was born during Cleveland's second term.
- Richard Folsom Cleveland (1897–1974) was born after his presidency.
- Francis Grover Cleveland (1903–1995) was the youngest, born in retirement.

Rise to Power

Cleveland's ascent to the presidency was remarkably swift, driven by his reputation for honesty and reform. His legal and political career in New York established him as a no-nonsense leader.

- Admitted to the bar in 1859, practiced law in Buffalo, and gained a reputation for diligence.

- Served as Erie County Sheriff (1870–1873), earning respect for his integrity and refusal to delegate executions.

- Elected Mayor of Buffalo in 1881, where he fought corruption and wasteful spending, earning the nickname "Veto Mayor."

- Became Governor of New York in 1882, continuing his reform agenda and gaining national attention.

- Nominated as the Democratic candidate for president in 1884, capitalizing on his anti-corruption credentials.

Influences

Cleveland's worldview was shaped by a combination of personal experiences, religious upbringing, and political philosophy, emphasizing limited government and personal responsibility.

- Presbyterian upbringing instilled a strong sense of duty and moral rectitude.

- Influenced by Democratic principles of the time, particularly Jeffersonian ideals of minimal federal intervention.

- Mentored by Buffalo lawyer Henry W. Rogers, who guided his early legal career.

- Inspired by the reform movements of the Gilded Age, they reacted against widespread political corruption.

Party Affiliation

Cleveland was a staunch Democrat, aligning with the party's conservative wing during an era of intense political partisanship.

- Identified with the Bourbon Democrats, who favored limited government, low taxes, and the gold standard.
- Opposed the Republican Party's protectionist tariffs and expansive federal policies.
- His commitment to Democratic principles often put him at odds with populist factions within his own party.

Presidency

Cleveland's two terms were defined by economic challenges, labor unrest, and his unwavering commitment to fiscal discipline. His presidencies tackled major issues like the Panic of 1893 and debates over currency, though his decisions were often polarizing.

- First term (1885–1889): Focused on reducing government spending, vetoing over 400 bills, including many private pension bills for Civil War veterans.
- Second term (1893–1897): Faced the Panic of 1893, a severe economic depression, and pushed for tariff reform and the gold standard.
- Sent federal troops to break the Pullman Strike of 1894, a controversial move that alienated labor unions.
- Advocated for civil service reform, expanding the merit-based system to reduce patronage.

- Dealt with foreign policy issues, including opposing the annexation of Hawaii and resolving disputes with Britain over Venezuela.

Accomplishments

Cleveland's presidency left a lasting mark on American governance, particularly in fiscal policy and government reform.

- Vetoed hundreds of bills to curb government spending, earning the nickname "Veto President."

- Signed the Interstate Commerce Act of 1887, the first federal regulation of railroads.

- Strengthened the civil service system, reducing political patronage through the Pendleton Act's expansion.

- Preserved the gold standard, stabilizing the economy during the Panic of 1893, though at significant political cost.

- Restored public lands by repealing fraudulent grants to railroads and other corporations.

First Lady's Contributions

Frances Folsom Cleveland, one of the youngest First Ladies, brought charm and modernity to the White House, influencing public perception and social causes.

- Hosted public receptions, making the White House more accessible to ordinary citizens.

- Supported charities, particularly for women and children, including the Washington Home for Friendless Women.
- Influenced fashion and social trends, becoming a national style icon.
- Advocated for education, supporting the establishment of kindergartens in Washington, D.C.
- Her poise and popularity softened Cleveland's stern public image, boosting his political appeal.

Positive Traits

Cleveland's personal strengths shaped his leadership style and earned him respect, though they sometimes limited his flexibility.

- Uncompromising integrity, refusing to engage in corrupt practices common in the Gilded Age.
- Strong work ethic, often personally reviewing legislation and government operations.
- Commitment to principle, prioritizing fiscal responsibility over political expediency.
- Courage in decision-making, such as vetoing popular but wasteful bills.

Negative Traits

Cleveland's stubbornness and rigidity often strained his relationships and limited his political effectiveness.

- Inflexibility, particularly on economic issues like the gold standard, alienated key constituencies like farmers and laborers.

- Aloof demeanor, which made him appear distant and unapproachable to the public.

- Reluctance to compromise, leading to conflicts with Congress and his own party.

- Limited vision for social reform, focusing narrowly on fiscal policy over broader societal issues.

Effects on Presidency

Cleveland's traits had a profound impact on his tenure, both enabling his successes and contributing to his challenges.

- His integrity restored public trust in government after years of corruption scandals.

- Stubborn adherence to the gold standard deepened economic woes for many Americans, fueling populist opposition.

- His aloofness hindered coalition-building, weakening his influence in Congress.

- Veto-heavy approach, while fiscally responsible, alienated veterans and other groups, costing him political capital.

Pets

Cleveland was not particularly known for keeping pets, but his household did include some animals, reflecting the era's simpler tastes.

- Owned a pet mockingbird during his first term, which he reportedly enjoyed listening to.
- His wife, Frances, kept a canary, a common pet in the White House at the time.
- No evidence of dogs or other larger animals, aligning with Cleveland's practical and unostentatious lifestyle.

Religious Persuasion

Cleveland's religious beliefs were rooted in his upbringing, but played a subdued role in his public life.

- Raised as a Presbyterian, influenced by his father's ministry.
- Maintained a private faith, rarely emphasizing religion in his political decisions.
- Attended Presbyterian services but avoided overt displays of religiosity, consistent with his reserved personality.

Interesting Anecdotes

Cleveland's life was filled with colorful stories that highlighted his character and the era's political climate.

- During the 1884 campaign, he faced a scandal over an alleged illegitimate child, responding candidly with, "Tell the truth," which mitigated political damage.
- As sheriff, he personally carried out two executions by hanging, a duty he found distasteful but refused to delegate.

- His wedding to Frances in the White House was a national sensation, with crowds gathering outside to celebrate.

- Known for working late into the night, he once said, "I have no right to be idle when the people's business is undone."

Ages at Death

Cleveland and his wife lived long lives after his presidency, retiring to Princeton, New Jersey.

- Grover Cleveland died at age 71 on June 24, 1908.
- Frances Folsom Cleveland died at age 83 on October 29, 1947.

Causes of Death

Both Cleveland and Frances died of natural causes, reflecting the medical realities of their times.

- Grover Cleveland succumbed to heart failure, likely exacerbated by his lifelong obesity and health issues.
- Frances Folsom Cleveland died of natural causes associated with old age.

Burial Locations

The couple was laid to rest in Princeton, where they had spent their retirement years.

- Grover Cleveland was buried in the Princeton Cemetery, Princeton, New Jersey.

- Frances Folsom Cleveland was also buried in the Princeton Cemetery, alongside her husband.

Conclusion

Grover Cleveland's legacy as the only president to serve non-consecutive terms reflects his enduring commitment to principle over popularity. His focus on fiscal conservatism, government reform, and integrity left a lasting impact, though his rigidity and economic policies stirred controversy. Frances Cleveland's grace and social contributions complemented his presidency, softening his stern image. Together, they navigated the challenges of a transformative era, leaving behind a complex but significant chapter in American history. Cleveland's dedication to "telling the truth" and governing responsibly remains a defining feature of his unique place in the nation's story.

William McKinley: Champion of American Prosperity and Expansion

Description

William McKinley, the 25th President of the United States, was a pivotal figure in American history, known for his leadership during a period of economic recovery and territorial expansion. A steadfast Republican, McKinley's presidency marked the United States' emergence as a global power, particularly through the Spanish-American War and the annexation of territories like Hawaii and the Philippines. His calm demeanor, commitment to protective tariffs, and tragic assassination cemented his legacy as a leader who guided the nation into a new era of influence.

Introduction

William McKinley served as president from March 4, 1897, until his assassination on September 14, 1901. His tenure was defined by economic policies that spurred industrial growth and foreign policy decisions that expanded American influence abroad. A man of deep religious faith and personal integrity, McKinley's leadership style was deliberate and consensus-driven, earning him both admiration and criticism. His assassination by an anarchist underscored the turbulent political climate of the era, leaving a lasting impact on American history.

Early Life

Born on January 29, 1843, in Niles, Ohio, William McKinley was the seventh of nine children in a middle-class family. His early life was shaped by modest circumstances and a strong work ethic, influenced by his parents' Scotch-Irish heritage and Presbyterian values.

- Grew up in a small town, attending local schools, and developing an early interest in debate and public speaking.
- Enrolled at Allegheny College in 1860 but left after one term due to illness and financial difficulties.
- Served in the Civil War as a young man, enlisting in the 23rd Ohio Volunteer Infantry at age 18, rising to the rank of brevet major by the war's end in 1865.
- His wartime experiences, including bravery at battles like Antietam, shaped his sense of duty and leadership.

Family

McKinley's family life centered around his close-knit upbringing and his devoted marriage to Ida Saxton McKinley.

- Parents: William McKinley Sr., a pig-iron manufacturer, and Nancy Allison McKinley, a devout Presbyterian who instilled strong moral values.

- Siblings: Eight siblings, with whom he maintained close ties; his family's support was crucial during his political career.

- Married to Ida Saxton in 1871, a union marked by deep affection but also tragedy due to the loss of their children and Ida's chronic health issues.

Children

William and Ida McKinley had two daughters, both of whom died young, profoundly affecting the couple.

- Katherine "Katie" McKinley (1871–1875): Died at age three of typhoid fever, a devastating loss for the McKinleys.

- Ida McKinley (1873–1873): Died in infancy at four months old, compounding the couple's grief.

- The loss of their children contributed to Ida's fragile health and McKinley's protective devotion to her.

Rise to Power

McKinley's political career began in Ohio and steadily progressed to national prominence through his legal and legislative work.

- Studied law after the Civil War and was admitted to the bar in 1867, practicing in Canton, Ohio.

- Elected prosecuting attorney of Stark County in 1869, marking his entry into politics.
- Served as a U.S. Congressman from Ohio (1877–1891, with interruptions), gaining recognition for his expertise on tariffs.
- Elected Governor of Ohio (1892–1896), where he focused on economic recovery and labor issues, boosting his national profile.
- Won the Republican presidential nomination in 1896, defeating Democrat William Jennings Bryan in a campaign centered on the gold standard and protective tariffs.

Influences

McKinley's worldview was shaped by key experiences and figures that guided his political philosophy.

- Civil War service under future president Rutherford B. Hayes, who became a mentor and instilled a sense of public service.
- Presbyterian faith, which emphasized duty, charity, and moral responsibility, influenced his personal and political decisions.
- Economic ideas of protectionism were inspired by the industrial growth of Ohio and the need to shield American industries from foreign competition.
- Close advisors like Mark Hanna, a political strategist who orchestrated McKinley's presidential campaigns and shaped his economic policies.

Party Affiliation

McKinley was a lifelong member of the Republican Party, aligning with its pro-business and expansionist policies.

- Advocated for protective tariffs to support American industry, a core Republican principle of the era.

- Supported the gold standard, opposing the populist push for free silver championed by Democrats like William Jennings Bryan.

- Embraced the party's emerging imperialist stance, leading to territorial acquisitions during his presidency.

Presidency

McKinley's presidency (1897–1901) was a transformative period, marked by economic recovery, the Spanish-American War, and global expansion.

- Assumed office during an economic depression, implementing policies to restore prosperity.

- Led the U.S. through the Spanish-American War (1898), resulting in victory and the acquisition of Puerto Rico, Guam, and the Philippines.

- Oversaw the annexation of Hawaii in 1898, expanding U.S. influence in the Pacific.

- Faced challenges in the Philippines, where an insurgency followed U.S. occupation, testing his administration's foreign policy.

- Re-elected in 1900, defeating Bryan again, but his second term was cut short by his assassination in 1901 at the Pan-American Exposition in Buffalo, New York.

Accomplishments

- Passed the Dingley Tariff Act (1897), raising tariffs to protect American industries and stimulate economic growth.
- Signed the Gold Standard Act (1900), stabilizing the U.S. currency by committing to gold-backed money.
- Led the U.S. to victory in the Spanish-American War, establishing America as a global power.
- Facilitated the annexation of Hawaii, strengthening U.S. strategic presence in the Pacific.
- Promoted industrial expansion, contributing to a period of economic prosperity known as the "McKinley Prosperity."

First Lady's Contributions

- Ida McKinley, despite chronic illness, served as First Lady with grace, though her contributions were limited by her health.
- Hosted White House social events when able, maintaining a dignified presence despite frequent seizures and frailty.
- Supported charitable causes, particularly those aiding women and children, though her involvement was often behind the scenes.

- Inspired public sympathy and awareness of health issues, as her condition was openly acknowledged, humanizing the first family.

Positive Traits

McKinley's personal strengths shaped his effective leadership and public image.

- Deliberate and methodical decision-making, fostering consensus among advisors and Congress.
- Deep empathy, particularly evident in his attentive care for Ida, endeared him to the public.
- Strong oratorical skills, honed in his early years, helped him articulate his vision during campaigns.
- Unwavering commitment to duty, rooted in his religious beliefs and wartime service.

Negative Traits

McKinley's weaknesses occasionally hindered his presidency.

- Reluctance to confront party factions directly sometimes led to compromises that diluted his policies.
- Over-reliance on advisors like Mark Hanna, which critics argued limited his independence.
- Initial hesitation on foreign policy, particularly regarding the Spanish-American War, delayed decisive action.

Effects on Presidency

McKinley's traits had a mixed impact on his leadership.

- His empathy and consensus-building fostered bipartisan support for his economic policies, aiding recovery.
- His cautious approach to foreign policy initially slowed responses to international crises but ultimately led to calculated decisions in the Spanish-American War.
- Dependence on advisors like Hanna streamlined his campaigns but sparked criticism of being overly influenced by party bosses.
- His devotion to Ida, while admirable, occasionally distracted him from political duties due to her frequent health crises.

Pets

The McKinleys were fond of animals, though their White House menagerie was modest.

- Kept a parrot named Washington Post, known for whistling patriotic tunes, which delighted visitors.
- Had several cats, which provided comfort to Ida during her illnesses.
- No dogs or other large animals were noted, likely due to the couple's focus on Ida's health and White House duties.

Religious Persuasion

McKinley's faith was a cornerstone of his life and leadership.

- Devout Methodist, having joined the church in his youth after a religious awakening.

- Regularly attended services at the Metropolitan Methodist Church in Washington, D.C.

- His faith influenced his moral outlook, emphasizing charity, humility, and public service, which guided his policies on labor and economic fairness.

Interesting Anecdotes

McKinley's life was filled with memorable stories that highlight his character.

- During the Civil War, as a young commissary sergeant, he braved enemy fire to deliver food and coffee to his regiment at Antietam, earning praise for his courage.

- He carried a red carnation in his lapel as a good-luck charm, a habit that became a public symbol; after his assassination, the carnation became Ohio's state flower.

- McKinley often personally escorted Ida to events, gently guiding her through crowds to manage her epilepsy discreetly, showcasing his devotion.

Age at Death

- William McKinley died at age 58 on September 14, 1901, eight days after being shot by anarchist Leon Czolgosz.

- Ida McKinley died at age 59 on May 26, 1907, outliving her husband by nearly six years.

Cause of Death

- William McKinley: Assassinated by two gunshot wounds to the abdomen; died of gangrene and infection due to inadequate medical treatment.
- Ida McKinley: Died of a stroke, compounded by her lifelong health struggles, including epilepsy and depression.

Burial Location

- William McKinley: Buried at the McKinley National Memorial in Canton, Ohio, a grand monument dedicated in 1907.
- Ida McKinley: Buried alongside her husband at the McKinley National Memorial in Canton, Ohio.

Conclusion

William McKinley's legacy as the "Champion of American Prosperity and Expansion" reflects his pivotal role in shaping the United States as an industrial and global power. His economic policies, including high tariffs and the gold standard, spurred growth, while his leadership in the Spanish-American War expanded America's reach. Despite personal tragedies and health challenges in his family, McKinley's empathy, faith, and deliberate leadership won him widespread respect. His assassination marked a tragic end to a transformative presidency, but his contributions laid the groundwork for the Progressive Era. Ida's quiet strength as First Lady, despite her frailty, complemented

McKinley's public service, leaving a poignant human legacy. Their story remains a testament to resilience, duty, and the complexities of leadership in a rapidly changing nation.

Theodore Roosevelt: Champion of the Progressive Era

Introduction

\Theodore Roosevelt, the 26th President of the United States, was a dynamic figure whose boundless energy, reformist zeal, and love for adventure shaped modern America. Known for his progressive policies, conservation efforts, and assertive foreign policy, Roosevelt left an indelible mark on the nation. His presidency from 1901 to 1909 bridged the 19th and 20th centuries, embodying a vision of vigorous leadership and national pride.

Early Life

Theodore Roosevelt was born into wealth and privilege but overcame personal challenges to become a larger-than-life figure. His early years were marked by intellectual curiosity and physical determination, setting the stage for his multifaceted career.

- Born on October 27, 1858, in New York City to a prominent family.
- Suffered from severe asthma as a child, which he combated with exercise and outdoor activities.
- Educated by private tutors and developed a passion for natural history and literature.
- Attended Harvard University (1876–1880), graduating magna cum laude with a degree in history.
- Studied law briefly at Columbia Law School but left to pursue public service.

Family

Roosevelt's family life was both a source of strength and tragedy, shaping his emotional resilience and public persona.

- Father: Theodore Roosevelt Sr., a philanthropist and businessman who inspired his son's sense of duty.
- Mother: Martha "Mittie" Bulloch, a Southern belle with Confederate sympathies.
- Siblings: Two sisters (Anna and Corinne) and one brother (Elliott).
- Married twice: First to Alice Hathaway Lee (1880–1884), then to Edith Kermit Carow (1886–1919).
- Maintained close ties with his extended family, who influenced his political and social views.

Children

Roosevelt was a devoted father whose children often accompanied him in his adventurous lifestyle.

- Alice Lee Roosevelt (1884–1980), from his first marriage, known for her wit and social prominence.

- Theodore Jr. (1887–1944) was a decorated soldier and public servant.

- Kermit (1889–1943) was an explorer and businessman who struggled with personal challenges.

- Ethel (1891–1977) was a philanthropist and advocate for social causes.

- Archibald (1894–1979) was a military officer and businessman.

- Quentin (1897–1918), a pilot killed in World War I, deeply affected Roosevelt.

Rise to Power

Roosevelt's ascent to the presidency was marked by bold actions and a knack for seizing opportunities.

- Began political career as a New York State Assemblyman (1882–1884), championing reform.

- Served as New York City Police Commissioner (1895–1897), rooting out corruption.

- Appointed Assistant Secretary of the Navy (1897–1898), where he prepared the Navy for the Spanish-American War.

- Gained fame as the leader of the Rough Riders during the Spanish-American War (1898).
- Elected Governor of New York (1899–1900), pushing progressive reforms.
- Became Vice President under William McKinley (1901) and assumed the presidency after McKinley's assassination.

Influences

Roosevelt's worldview was shaped by a blend of intellectual, cultural, and personal forces.

- Inspired by his father's emphasis on civic duty and moral integrity.
- Influenced by naturalists like John James Audubon, fueling his passion for conservation.
- Admired Abraham Lincoln's leadership and commitment to national unity.
- Shaped by the Social Darwinism of his era, believing in competition and national strength.
- Drew from his travels and experiences as a rancher in the Dakota Territory.

Party Affiliation

Roosevelt's political alignment evolved, reflecting his commitment to reform over strict party loyalty.

- Initially, a staunch Republican, aligning with the party's progressive wing.

- Broke with conservative Republicans to form the Progressive ("Bull Moose") Party in 1912.
- Advocated for government intervention to curb corporate excesses and protect workers.
- Returned to the Republican fold later but remained a progressive at heart.

Presidency

Roosevelt's presidency (1901–1909) was defined by bold domestic reforms and an assertive foreign policy.

- Assumed office on September 14, 1901, after McKinley's assassination.
- Promoted the "Square Deal," focusing on consumer protection, corporate regulation, and conservation.
- Strengthened the executive branch, using the "bully pulpit" to advocate for change.
- Expanded U.S. global influence, particularly through the Panama Canal project.
- Mediated the Russo-Japanese War, earning the Nobel Peace Prize in 1906.
- Left office in 1909, handpicking William Howard Taft as his successor.

Accomplishments

- Established the National Park System, preserving millions of acres for future generations.

- Signed the Pure Food and Drug Act (1906) and the Meat Inspection Act (1906) to protect consumers.
- Broke up monopolies, earning the nickname "trust-buster."
- Facilitated the construction of the Panama Canal, enhancing U.S. global trade.
- Strengthened the U.S. Navy, projecting American power worldwide.
- Won the Nobel Peace Prize for negotiating peace in the Russo-Japanese War.

First Lady's Contributions

Edith Roosevelt, Theodore's second wife, was a poised and influential First Lady who modernized the role.

- Oversaw the renovation of the White House, creating the West Wing for presidential offices.
- Established the role of a formal First Lady's staff to manage social events.
- Promoted cultural events, hosting musical performances and literary gatherings.
- Advocated for historic preservation, supporting efforts to maintain national landmarks.
- Managed the Roosevelt family's public image with grace and discretion.

Positive Traits

Roosevelt's strengths amplified his effectiveness as a leader but sometimes led to overreach.

- Energetic and charismatic, inspiring public confidence and action.
- Visionary, with a forward-thinking approach to conservation and reform.
- Courageous, both physically (as a war hero) and politically (challenging powerful interests).
- Intellectual, with a deep knowledge of history and science that informed his policies.
- Positive traits strengthened his ability to rally support for progressive causes and expand presidential power.

Negative Traits

Roosevelt's flaws occasionally undermined his goals and strained relationships.

- Impulsive, sometimes acting without fully considering consequences.
- Overly aggressive in foreign policy, risking diplomatic tensions.
- Stubborn, clashing with allies over ideological differences.
- Tendency to dominate conversations and decisions, alienating some advisors.

- Negative traits led to occasional diplomatic missteps and political feuds, notably his split with the Republican Party.

Pets

The Roosevelt White House was a lively menagerie, reflecting the family's love for animals.

- Bleistein, a horse used by Roosevelt for riding.
- Pete, a bull terrier who famously bit a White House guest.
- Skip, a mongrel dog beloved by the family.
- Josiah, a badger given as a gift during a Western tour.
- Manchu, a Pekingese dog given to Alice by the Empress of China.
- Various snakes, guinea pigs, and a bear cub delight the Roosevelt children.

Religious Persuasion

Roosevelt's faith influenced his moral outlook but was not rigidly dogmatic.

- Raised in the Dutch Reformed Church, attending services regularly.
- Believed in a practical Christianity that emphasized good works and social justice.
- Rarely discussed personal faith publicly, focusing instead on universal ethical principles.

- Supported religious tolerance, welcoming diverse faiths in his administration.

Interesting Anecdotes

Roosevelt's colorful life was filled with memorable stories that showcased his personality.

- Once delivered a 90-minute speech after being shot in the chest during his 1912 campaign.
- Invited African American leader Booker T. Washington to dine at the White House, sparking controversy.
- Wrestled with his children in the White House, turning formal spaces into playgrounds.
- Led expeditions in the Amazon after his presidency, nearly dying from infection.
- Coined the phrase "speak softly and carry a big stick" to describe his foreign policy.

Ages at Death

- Theodore Roosevelt died at age 60 on January 6, 1919.
- Edith Roosevelt died at age 87 on September 30, 1948.

Causes of Death

- Theodore Roosevelt succumbed to a coronary occlusion, exacerbated by lingering health issues from his Amazon expedition.
- Edith Roosevelt died of natural causes related to old age.

Burial Locations

- Theodore Roosevelt was buried at Youngs Memorial Cemetery in Oyster Bay, New York.
- Edith Roosevelt was buried beside him at Youngs Memorial Cemetery.

Conclusion

Theodore Roosevelt's legacy as the "Champion of the Progressive Era" endures through his transformative policies, conservation achievements, and bold leadership. His presidency redefined the role of the federal government in protecting citizens and natural resources while projecting American strength abroad. Despite personal and political flaws, Roosevelt's relentless drive and vision left a lasting impact on the nation. Edith Roosevelt complemented his legacy with her quiet strength and modernization of the First Lady's role, together shaping an era of progress and possibility.

William Howard Taft: The Trust-Busting President

Introduction

William Howard Taft, the 27th President of the United States, served from 1909 to 1913 and is most remembered for his trust-busting efforts to regulate big business. A man of judicial temperament, Taft's career spanned law, governance, and diplomacy, culminating in his historic appointment as the 10th Chief Justice of the United States Supreme Court after his presidency. His legacy reflects a commitment to legal reform, progressive policies, and international arbitration, though his presidency faced challenges due to political divisions within his party.

Early Life

William Howard Taft was born on September 15, 1857, in Cincinnati, Ohio, into a prominent family with a strong legal and political background. His early years shaped his disciplined work ethic and intellectual rigor.

- Grew up in Cincinnati's affluent Mount Auburn neighborhood.
- Excelled academically at Woodward High School, graduating second in his class.
- Attended Yale University, earning a bachelor's degree in 1878, and later studied law at Cincinnati Law School, admitted to the Ohio bar in 1880.
- Developed a passion for law and public service, influenced by his father's career as a judge and politician.

Family

Taft's family played a significant role in his personal and professional life, providing support and shaping his values.

- Father: Alphonso Taft, a distinguished attorney, judge, and U.S. Attorney General under President Ulysses S. Grant.
- Mother: Louisa Maria Torrey, a nurturing figure who encouraged education and civic duty.
- Siblings: Taft was one of six children, with brothers who also pursued successful careers in law and business.

- Married Helen "Nellie" Herron in 1886, forming a partnership that influenced his political and social life.

Children

Taft and Nellie had three children, each of whom achieved notable success, reflecting the family's emphasis on education and public service.

- Robert Alphonso Taft (1889–1953): Became a prominent U.S. Senator from Ohio, known as "Mr. Republican" for his conservative leadership.

- Helen Herron Taft (1891–1987): An academic who earned a Ph.D. and served as a professor and dean at Bryn Mawr College.

- Charles Phelps Taft II (1897–1983): A lawyer and mayor of Cincinnati, active in local politics and philanthropy.

Rise to Power

Taft's ascent to the presidency was marked by a steady progression through legal and administrative roles, bolstered by his association with key Republican figures.

- Began as a lawyer in Cincinnati, then served as Assistant Prosecutor of Hamilton County (1881–1883).

- Appointed Assistant Solicitor of Hamilton County (1885) and Superior Court judge (1887).

- Named U.S. Solicitor General by President Benjamin Harrison in 1890, arguing cases before the Supreme Court.

- Served as a federal circuit judge (1892–1900), gaining a reputation for fairness and legal acumen.

- Appointed Governor-General of the Philippines (1901–1904) by President William McKinley, where he implemented reforms to stabilize the region.

- Served as Secretary of War under President Theodore Roosevelt (1904–1908), overseeing the Panama Canal project and earning Roosevelt's trust.

- Nominated as the Republican candidate for president in 1908, largely due to Roosevelt's endorsement.

Influences

Taft's worldview and policies were shaped by mentors, family, and his legal background.

- Theodore Roosevelt: His predecessor and friend, whose progressive policies initially guided Taft's administration.

- Father Alphonso Taft: Instilled a respect for law and public service.

- Nellie Taft: Encouraged his political ambitions and influenced his social policies.

- Legal training: Emphasized judicial restraint and constitutional adherence, which defined his approach to governance.

Party Affiliation

Taft was a lifelong Republican, aligned with the party's conservative and progressive factions, though his presidency leaned toward conservatism.

- Supported Republican principles of economic growth, limited government, and strong national defense.
- Struggled to balance the progressive wing, led by Roosevelt, with the conservative "Old Guard" Republicans.
- His conservative leanings led to a split in the party, contributing to his 1912 election loss.

Presidency

Taft's presidency (1909–1913) focused on antitrust enforcement, tariff reform, and international diplomacy, though it was marred by political conflicts.

- Pursued aggressive antitrust policies, breaking up monopolies like Standard Oil and American Tobacco.
- Signed the Payne-Aldrich Tariff Act (1909), which aimed to lower tariffs but disappointed progressives due to compromises.
- Established the Department of Labor and supported the 16th Amendment, enabling a federal income tax.
- Promoted "Dollar Diplomacy," using economic influence to expand U.S. interests in Latin America and Asia.

- Faced challenges from the Progressive movement, leading to a rift with Roosevelt and the formation of the Progressive Party.
- Lost the 1912 election to Woodrow Wilson due to the Republican Party's split.

Accomplishments

- Initiated 80 antitrust lawsuits, earning the "trust-buster" moniker.
- Established the U.S. Postal Savings System to provide banking for underserved communities.
- Created the Department of Labor to address workers' rights and labor issues.
- Supported the 16th Amendment, ratified in 1913, establishing the federal income tax.
- Strengthened U.S. foreign policy through Dollar Diplomacy, fostering economic ties abroad.
- Appointed six Supreme Court justices, shaping the judiciary for decades.
- Became Chief Justice of the United States (1921–1930), the only president to serve in this role.

First Lady's Contributions

- Planted the first cherry blossom trees in Washington, D.C., a gift from Japan, creating a lasting cultural landmark.
- Transformed the White House into a social hub, hosting musical performances and diplomatic events.

- Advocated for public health, supporting safe working conditions for federal employees.
- Influenced White House renovations, enhancing its aesthetic and functionality.
- Promoted cultural diplomacy, strengthening U.S.-Japan relations through her cherry blossom initiative.

Positive Traits

Taft's personal qualities shaped his presidency, often aligning with his judicial and administrative strengths.

- Analytical: His legal mind ensured thorough policy decisions, particularly in antitrust cases.
- Diligent: Worked tirelessly on judicial and administrative reforms, earning respect for his work ethic.
- Diplomatic: Fostered international goodwill through arbitration treaties and Dollar Diplomacy.
- Effect on presidency: His methodical approach strengthened legal and economic policies but sometimes slowed decision-making, frustrating allies.

Negative Traits

Taft's weaknesses contributed to political challenges during his term.

- Indecisive: Hesitancy in addressing party conflicts alienated progressives.
- Reserved: Lacked the charisma of Roosevelt, making it harder to rally public support.

- Effect on presidency: His reluctance to engage in political maneuvering led to the Republican Party's split and his 1912 defeat.

Pets

The Tafts brought a unique pet to the White House, reflecting their family-oriented lifestyle.

- Carow, a Holstein cow, provided fresh milk for the White House, a practical addition during their tenure.
- The cow was a gift from a Wisconsin senator, symbolizing agricultural support.

Religious Persuasion

\Taft's faith influenced his moral outlook, but was not a public focal point.

- Raised as a Unitarian, emphasizing reason and individual conscience.
- Attended Unitarian services but kept his religious beliefs private, focusing on secular governance.
- His faith reinforced his commitment to fairness and ethical leadership.

Interesting Anecdotes

Taft's life was marked by colorful stories that highlight his personality and era.

- Known for his large stature, Taft once got stuck in a White House bathtub, leading to the installation of a larger one.

- As Governor-General of the Philippines, he declined a lucrative job offer to stay in public service, showing his dedication.
- Nellie Taft suffered a stroke early in his presidency (1909), yet she continued hosting White House events with remarkable resilience.
- Taft's love for baseball led him to throw the first presidential pitch at a Major League game in 1910, starting a tradition.

Ages at Death

- William Howard Taft: Died at age 72 on March 8, 1930.
- Helen "Nellie" Taft: Died at age 81 on May 22, 1943.

Causes of Death

- William Howard Taft: Died of heart disease, exacerbated by years of poor health and obesity.
- Helen "Nellie" Taft: Died of natural causes related to old age, having outlived her husband by 13 years.

Burial Locations

- William Howard Taft: Buried at Arlington National Cemetery, Virginia, one of only two presidents interred there.
- Helen "Nellie" Taft: Also buried at Arlington National Cemetery, alongside her husband.

Conclusion

William Howard Taft's legacy as the trust-busting president reflects his commitment to curbing monopolies and strengthening the rule of law. His judicial mindset and progressive reforms, such as antitrust enforcement and the income tax, left a lasting impact on American governance. Despite political missteps and a reserved demeanor that cost him re-election, Taft's post-presidency role as Chief Justice cemented his dedication to justice. Nellie Taft's cultural contributions, notably the cherry blossoms, enhanced the White House's legacy. Together, their partnership exemplified public service, leaving a multifaceted mark on American history.

Woodrow Wilson: Architect of the League of Nations

Description

Thomas Woodrow Wilson, the 28th President of the United States, was a scholar, reformer, and progressive leader who guided the nation through World War I and championed the League of Nations. A complex figure, his presidency was marked by significant domestic reforms and controversial policies, leaving a lasting impact on American and global politics.

Introduction

Woodrow Wilson served as president from 1913 to 1921, navigating the United States through a transformative period in domestic and international affairs. Known for his intellectual rigor and progressive ideals, he introduced sweeping reforms while facing criticism for his handling of civil liberties and racial issues. His vision for global peace through the League of Nations remains his most enduring legacy, though his presidency was also shaped by personal health struggles and polarizing decisions.

Early Life

Born on December 28, 1856, in Staunton, Virginia, Woodrow Wilson grew up in the South during the Civil War and Reconstruction. His early experiences in a region scarred by conflict shaped his views on governance and unity.

- Raised in a devout Presbyterian family, with his father, Joseph Wilson, a minister.
- Educated at home before attending Davidson College and later Princeton University (1879).
- Studied law at the University of Virginia and briefly practiced before earning a Ph.D. in political science from Johns Hopkins University (1886).
- His academic background set him apart as one of the most educated presidents in U.S. history.

Family

Wilson's family life was marked by deep personal connections and tragedy, with two marriages shaping his personal and public life.

- Parents: Joseph Ruggles Wilson and Jessie Janet Woodrow, both of Scottish descent.
- First wife: Ellen Axson Wilson, married in 1885, died in 1914 during his presidency.

- Second wife: Edith Bolling Galt, married in 1915, played a significant role during his later presidency.
- Siblings: Wilson had three siblings—Marion, Anne, and Joseph—though he remained closest to his parents.

Children

Wilson had three daughters from his first marriage to Ellen Axson Wilson, all of whom were active in their own ways but maintained relatively private lives.

- Margaret Woodrow Wilson (1886–1944): Never married, pursued a career in music, and later became a missionary in India.
- Jessie Woodrow Wilson Sayre (1887–1933): Married Francis Bowes Sayre, active in social reform and women's suffrage.
- Eleanor Randolph Wilson McAdoo (1889–1967): Married William Gibbs McAdoo, Wilson's Treasury Secretary, and was involved in public life.

Rise to Power

Wilson's ascent to the presidency was rapid, transitioning from academia to politics in a few short years.

- Served as president of Princeton University (1902–1910), implementing progressive educational reforms.
- Elected governor of New Jersey (1910–1913), where he gained national attention for combating political corruption and passing reform legislation.

- Nominated as the Democratic candidate for president in 1912, capitalizing on a split in the Republican Party.
- Won the presidency in 1912 against incumbent William Howard Taft and Theodore Roosevelt, securing 435 electoral votes.

Influences

Wilson's worldview was shaped by a blend of intellectual, religious, and political influences.

- Presbyterian upbringing instilled a sense of moral duty and divine purpose in governance.
- Admired British parliamentary systems, particularly the oratory of William Gladstone.
- Influenced by progressive thinkers like Walter Bagehot and his own studies in political science, emphasizing strong executive leadership.
- The Civil War and Reconstruction informed his belief in national unity and centralized governance.

Party Affiliation

Wilson was a lifelong Democrat, aligning with the party's progressive wing during his political career.

- Championed the "New Freedom" platform, advocating for economic competition, individual liberty, and limited government intervention.
- Broke with conservative Democrats by supporting labor rights and regulatory reforms.

- His presidency shifted the Democratic Party toward progressive policies, influencing its trajectory for decades.

Presidency

Wilson's presidency (1913–1921) was defined by progressive domestic reforms, World War I leadership, and a vision for global cooperation, though marred by controversies over civil rights and wartime policies.

- The first term focused on economic reforms, including tariff reduction and banking restructuring.
- Led the U.S. into World War I in 1917, shifting from neutrality to active involvement.
- Proposed the Fourteen Points, a blueprint for post-war peace, including the League of Nations.
- Faced criticism for suppressing dissent through the Espionage and Sedition Acts and for his administration's racial segregation policies.
- Suffered a debilitating stroke in 1919, limiting his effectiveness in his final years.

Accomplishments

- Established the Federal Reserve System (1913), creating a centralized banking system.
- Passed the Clayton Antitrust Act (1914), strengthening anti-monopoly laws.
- Signed the Federal Trade Commission Act (1914), promoting fair competition.

- Advocated for the 19th Amendment, granting women's suffrage (1920).
- Proposed the League of Nations, earning the Nobel Peace Prize (1919).
- Oversaw U.S. victory in World War I, shaping post-war global order.

First Lady's Contributions

- **Ellen Axson Wilson:**
- Promoted housing reform for Washington, D.C.'s impoverished areas.
- Supported the arts, hosting cultural events at the White House.
- Advocated for better living conditions for federal employees.

- **Edith Bolling Wilson:**
- Managed access to Wilson after his 1919 stroke, effectively controlling communication with the president.
- Supported war bond drives during World War I.
- Preserved Wilson's legacy through her memoirs and historical efforts.

Positive Traits and Effects

Wilson's intellectual rigor, eloquence, and commitment to reform drove significant legislative successes.

- His academic background informed policies like the Federal Reserve, strengthening the economy.
- His principled leadership inspired progressive reforms and rallied the nation during wartime.
- His vision for international cooperation laid the groundwork for modern global institutions.
Negative traits, such as inflexibility and moral absolutism, hindered his presidency.
- His refusal to compromise on the League of Nations led to its rejection by the U.S. Senate.
- His dismissive attitude toward civil rights exacerbated racial tensions, alienating key constituencies.
- His authoritarian streak during wartime damaged civil liberties, drawing lasting criticism.

Pets

Wilson maintained a relatively modest household but was known to have animals at the White House.

- Kept a flock of sheep on the White House lawn to graze and promote wartime wool production.
- Old Ike, a tobacco-chewing ram, was a favorite among the sheep.
- Owned a cat named Puffins, which roamed the White House during his presidency.

Religious Persuasion

Wilson's deep Presbyterian faith profoundly influenced his worldview and leadership style.

- Believed in a divine mission to improve society and promote peace.
- Regularly attended church services and incorporated moral principles into his speeches.
- His faith shaped his idealism, particularly in crafting the Fourteen Points and the League of Nations.

Interesting Anecdotes

- As Princeton's president, Wilson once locked horns with faculty over campus dining clubs, showing his stubborn streak.
- During World War I, he allowed sheep to graze on the White House lawn, both for wool and to cut maintenance costs.
- After his stroke, Edith Wilson effectively acted as a gatekeeper, leading some to call her the "first female president."
- Wilson was an avid baseball fan, attending games and even throwing out the first pitch at professional matches.

Ages at Death, Causes of Death, and Burial Locations

- **Woodrow Wilson:** Died at age 67 on February 3, 1924, from complications of a stroke and heart issues. Buried at the Washington National Cathedral, Washington, D.C., the only president interred there.
- **Ellen Axson Wilson:** Died at age 54 on August 6, 1914, from Bright's disease (kidney failure). Buried at Myrtle Hill Cemetery, Rome, Georgia.

- **Edith Bolling Wilson**: Died at age 89 on December 28, 1961, from heart failure. Buried alongside Woodrow Wilson at the Washington National Cathedral.

Conclusion

Woodrow Wilson's legacy as the architect of the League of Nations reflects his ambitious vision for global peace, even as his presidency was marked by contradictions. His progressive reforms transformed the American economy, but his failures on civil rights and civil liberties cast a shadow. A scholar-turned-president, Wilson's intellectualism and idealism drove historic achievements, yet his inflexibility and health struggles limited his effectiveness. His life, shaped by faith, family, and a commitment to reform, remains a pivotal chapter in American history, remembered for both its triumphs and its flaws.

Warren G. Harding: The Teapot Dome Scandal President

Description

Warren Gamaliel Harding, the 29th President of the United States, served from 1921 until his death in 1923. Known for his affable personality and handsome appearance, Harding's presidency was marred by corruption scandals, most notably the Teapot Dome affair, which overshadowed his domestic and foreign policy efforts. His administration promised a "return to normalcy" after World War I, but his legacy is largely defined by the misconduct of his appointees.

Introduction

Harding's life reflects the complexities of early 20th-century American politics. Born in rural Ohio, he rose from a small-town newspaper editor to a U.S. senator and president. His charm and political connections propelled him to the White House, but his lack of strong leadership allowed corrupt associates to exploit his administration.

Despite his short tenure, Harding's presidency and the scandals that followed shaped public distrust in government.

Early Life

Harding was born on November 2, 1865, in Blooming Grove, Ohio, to George Tryon Harding and Phoebe Elizabeth Dickerson. Raised in a modest farming community, he developed an early interest in journalism and public speaking.

- Attended Ohio Central College from 1879 to 1882, where he honed his oratorical skills.
- Worked odd jobs, including teaching and selling insurance, before entering the newspaper business.
- Purchased the struggling *Marion Star* in 1884, turning it into a successful local paper.

Family

Harding's family was rooted in Ohio, with a mix of farmers and small-town professionals. His father, George, was a homeopathic physician, and his mother, Phoebe, a midwife.

- Married Florence Kling DeWolfe, a divorcee five years his senior, on July 8, 1891.
- No siblings played a prominent role in his public life, though he maintained close ties with his extended family.

- Florence, known as "The Duchess," was a dominant influence, managing both Harding's personal and political affairs.

Children

Harding had no children with Florence. However, he fathered an illegitimate daughter, Elizabeth Ann Blaesing, with his mistress, Nan Britton.

- Elizabeth was born in 1919, and Harding provided financial support but never publicly acknowledged her.
- The affair remained a secret until Britton's 1927 book, *The President's Daughter*, revealed it, causing public controversy.

Rise to Power

Harding's political career began in Ohio, leveraging his newspaper and social connections. His affability and moderate views made him a compromise candidate for higher office.

- Elected to the Ohio State Senate in 1899, serving two terms.
- Served as Ohio's lieutenant governor from 1904 to 1906.
- Won a U.S. Senate seat in 1914, gaining national visibility.
- Nominated as the Republican presidential candidate in 1920, chosen in a "smoke-filled room" deal at the convention.

- Won the presidency in a landslide against Democrat James M. Cox, promising a "return to normalcy."

Influences

Harding's political philosophy was shaped by his small-town upbringing and business-oriented associates.

- Admired William McKinley, Ohio's former president, for his pro-business policies.
- Influenced by Republican Party bosses like Harry Daugherty, who managed his campaigns.
- His wife, Florence, guided his career decisions, pushing him toward ambitious goals.
- His journalistic background fostered a belief in open communication, though he avoided confrontational leadership.

Party Affiliation

Harding was a lifelong Republican, aligning with the party's conservative, pro-business wing.

- Supported high tariffs, tax cuts, and limited government intervention in the economy.
- Advocated for Republican unity, often mediating between progressive and conservative factions.
- His moderate stance appealed to voters weary of Progressive Era reforms and World War I activism.

Presidency

Harding's presidency (March 4, 1921–August 2, 1923) focused on economic recovery and international disengagement. Scandals, however, defined his tenure.

- Appointed a mix of capable leaders (e.g., Herbert Hoover) and corrupt cronies (e.g., Albert Fall).
- Signed the Budget and Accounting Act of 1921, creating the Bureau of the Budget.
- Hosted the Washington Naval Conference (1921–1922), leading to arms limitation treaties.
- The Teapot Dome scandal involved illegal leasing of federal oil reserves, exposed after his death.
- Died suddenly during a cross-country tour, sparing him from facing the scandals' full fallout.

Accomplishments

- Established the Bureau of the Budget, modernizing federal financial management.
- Signed the Sheppard-Towner Act, providing federal aid for maternal and child health.
- Secured naval disarmament agreements at the Washington Naval Conference.
- Pardoned socialist Eugene V. Debs, imprisoned during World War I for anti-war activism.
- Promoted economic growth through tax cuts and reduced government spending.

First Lady's Contributions

- Hosted White House events to boost Harding's public image and support veterans.
- Advocated for women's involvement in politics, reflecting her own influence.
- Supported the humane treatment of animals, backing early animal welfare initiatives.
- Visited hospitals and orphanages, promoting public health and child welfare.
- Managed Harding's schedule and correspondence, acting as his de facto chief of staff.

Positive Traits and Effects

Harding's charm and conciliatory nature helped him navigate political divisions.

- His likability won allies in Congress, easing passage of his economic agenda.
- His calm demeanor reassured a nation fatigued by war and reform.
- His trust in advisors allowed competent officials like Hoover to excel.
- These traits fostered short-term stability but left him vulnerable to manipulation by dishonest associates.

Negative Traits and Effects

Harding's weaknesses as a leader contributed to his administration's failures.

- His indecisiveness and reluctance to confront allies enabled corruption.
- His loyalty to friends led to the appointment of unqualified cronies to key posts.
- His extramarital affairs, including with Nan Britton, risked public scandal.
- These flaws undermined his presidency's integrity, as scandals like Teapot Dome eroded public confidence.

Pets

The Hardings were fond of animals, and their pets were popular with the public.

- Laddie Boy, an Airedale terrier, was a White House favorite, often featured in news stories.
- Laddie Boy had his own chair at cabinet meetings, symbolizing the Hardings' approachable image.
- The couple also kept canaries and other small pets in the White House.

Religious Persuasion

Harding was raised in a Baptist household but was not overtly religious in public life.

- Attended church sporadically, aligning with mainstream Protestant values.
- His speeches occasionally referenced Christian principles, reflecting cultural norms.

- Florence shared similar views, supporting charitable causes tied to religious organizations.

Interesting Anecdotes

Harding's life included quirky and revealing moments.

- He once lost a set of White House china in a poker game with friends.
- As a newspaper editor, he wrote glowing editorials about himself in the third person.
- During his 1923 western tour, he shook hands with so many people that his hand swelled painfully.
- Rumors persisted that Florence poisoned him to avoid scandal, though no evidence supports this.

Ages at Death, Causes, and Burial Locations

- **Warren G. Harding**: Died at age 57 on August 2, 1923, likely of a heart attack, in San Francisco. Buried in the Harding Tomb, Marion, Ohio.
- **Florence Harding**: Died at age 64 on November 21, 1924, of kidney failure, in Marion, Ohio. Buried in the Harding Tomb, Marion, Ohio.

Conclusion

Warren G. Harding's presidency, though brief, encapsulates the contradictions of the early 1920s—a desire for stability clashing with unchecked corruption. His charm and good intentions were overshadowed by poor judgment in choosing associates, leading to scandals like Teapot Dome that tainted his legacy. Florence's ambition and contributions as First Lady provided a glimpse of potential, but Harding's weaknesses as a leader defined his tenure. His sudden death left unanswered questions, cementing his place as a cautionary figure in American political history.

Calvin Coolidge: The Silent Leader of Prosperity

Introduction

Calvin Coolidge, the 30th President of the United States, served from 1923 to 1929, a period marked by economic prosperity and minimal government intervention. Known as "Silent Cal" for his reserved demeanor, Coolidge's presidency embodied fiscal conservatism and a belief in limited government. His leadership during the Roaring Twenties shaped an era of industrial growth and social change, though his policies also set the stage for the economic challenges that followed. This summary explores his life, legacy, and the contributions of his wife, Grace Coolidge, providing a comprehensive look at their personal and public lives.

Early Life

John Calvin Coolidge Jr. was born on July 4, 1872, in Plymouth Notch, Vermont, a small rural community. Raised in a modest, hardworking environment, his early experiences instilled values of thrift, duty, and simplicity that defined his character and political philosophy.

- Grew up in a farming family, helping with chores like tending livestock and harvesting crops.
- Attended local schools, showing aptitude in academics, particularly in history and public speaking.
- Enrolled at Amherst College in 1891, graduating cum laude in 1895 with a strong foundation in law and political thought.
- Studied law in Northampton, Massachusetts, and was admitted to the bar in 1897, beginning a legal career that laid the groundwork for his political ascent.

Family

Coolidge's family life was anchored by strong New England roots and a deep commitment to personal values. His upbringing and marriage reflected his preference for stability and tradition.

- Father, John Calvin Coolidge Sr., was a farmer, storekeeper, and local official who instilled discipline and civic duty.
- Mother, Victoria Josephine Moor, died when Coolidge was 12, profoundly affecting his emotional reserve.

- Sister, Abigail Grace Coolidge, died at 15 from appendicitis, further shaping his stoic outlook.
- Married Grace Anna Goodhue on October 4, 1905, in Burlington, Vermont, forming a partnership that balanced his quiet nature with her warmth.

Children

Calvin and Grace Coolidge had two sons, whose lives were touched by both privilege and tragedy during their time in the public eye.

- John Coolidge (1906–2000): Studied at Amherst College, worked in business, and lived a private life after his father's presidency. He married Florence Trumbull, daughter of Connecticut Governor John H. Trumbull.
- Calvin Coolidge Jr. (1908–1924): Died at 16 from blood poisoning caused by a blister from playing tennis. His death deeply affected the Coolidge family, with the president reportedly becoming more withdrawn.

Rise to Power

Coolidge's political career began in local Massachusetts politics, marked by steady progression and a reputation for integrity and pragmatism.

- Elected to the Northampton City Council in 1898, starting a series of local roles, including city solicitor.
- Served in the Massachusetts House of Representatives (1907–1908) and Senate (1912–1915), advocating for fiscal restraint.

- Became Massachusetts Governor in 1916, gaining national attention for his handling of the 1919 Boston Police Strike, declaring, "There is no right to strike against the public safety."
  - Elected Vice President in 1920 under Warren G. Harding, assuming the presidency on August 2, 1923, after Harding's sudden death.

Influences

Coolidge's philosophy was shaped by personal experiences, historical figures, and economic thinkers who reinforced his belief in limited government and individual responsibility.

- Admired the writings of economist William Graham Sumner, who emphasized laissez-faire economics.

- Influenced by his father's example of public service and frugality in rural Vermont.

- Drew inspiration from Abraham Lincoln's emphasis on unity and constitutional principles.

- Shaped by the Progressive Era's debates, though he leaned toward traditionalism over reform.

Party Affiliation

Coolidge was a steadfast member of the Republican Party, aligning with its pro-business and conservative factions.

- Supported Republican principles of low taxes, reduced government spending, and minimal regulation.

- Opposed Progressive-era reforms that expanded federal power, favoring state and local governance.
- His party loyalty helped him secure nominations, though his reserved style sometimes frustrated party leaders seeking more dynamic leadership.

Presidency

Coolidge's presidency, from 1923 to 1929, was defined by economic prosperity, known as the "Coolidge Prosperity," and a hands-off approach to governance. He focused on reducing federal debt, cutting taxes, and promoting business growth.

- Assumed office after Harding's death, restoring public trust amid scandals like Teapot Dome.
- Signed the Revenue Acts of 1924, 1926, and 1928, reducing income and corporate taxes to stimulate economic growth.
- Vetoed bills expanding federal spending, including farm subsidies, to maintain budget surpluses.
- Promoted foreign policy restraint, supporting the Kellogg-Briand Pact (1928) to outlaw war.
- Declined to run for re-election in 1928, leaving office with high approval but warnings of economic overconfidence.

Accomplishments

Coolidge's presidency left a lasting impact through policies that fueled economic growth and maintained stability, though some decisions had mixed long-term effects.

- Reduced national debt by $2 billion through strict budget discipline.

- Lowered federal income tax rates, boosting consumer spending and economic expansion.

- Signed the Radio Act of 1927, establishing federal regulation of radio broadcasting.

- Strengthened Native American rights by signing the Indian Citizenship Act of 1924, which granted citizenship to all Native Americans born in the U.S.

- Maintained peace and prosperity, fostering the Roaring Twenties' cultural and economic vibrancy.

First Lady's Contributions

Grace Coolidge, known for her warmth and charm, complemented her husband's reserved nature and made significant contributions as First Lady.

- Advocated for the deaf, supporting organizations like the American School for the Deaf, reflecting her background as a teacher.

- Hosted White House events, enhancing its social role and restoring public confidence after the Harding scandals.

- Supported veterans' causes, visited hospitals, and promoting welfare for World War I soldiers.

- Championed animal welfare, particularly through her affection for the family's pets, endearing her to the public.

- Strengthened the First Lady's role as a public figure, balancing grace with advocacy.

Positive Traits

Coolidge's personal qualities shaped a presidency admired for its stability and restraint.

- Integrity: His honesty restored trust in government after the Harding-era scandals.
- Discipline: Strict adherence to fiscal conservatism reduced federal spending and debt.
- Decisiveness: His handling of the Boston Police Strike demonstrated firm leadership in crises.
- Humility: Lived modestly, reflecting his Vermont roots, which resonated with many Americans.

Negative Traits

Coolidge's traits, while effective in his era, also had drawbacks that influenced his presidency's legacy.

- Aloofness: His reserved nature limited public engagement and party coordination.
- Overreliance on laissez-faire: His minimal regulation contributed to speculative excesses, foreshadowing the 1929 stock market crash.
- Limited vision: Resistance to federal intervention hindered proactive solutions for emerging economic and social issues.

- Emotional restraint: Personal tragedies, like his son's death, deepened his detachment, affecting empathy and engagement with national hardships.

Pets

The Coolidges were known for their affection for animals, keeping a variety of pets in the White House that added charm to their public image.

- Had multiple dogs, including Rob Roy (a collie) and Prudence Prim (a white terrier).
- Owned a raccoon named Rebecca, a gift intended for Thanksgiving dinner but kept as a pet.
- Kept cats, canaries, and a mockingbird, reflecting Grace's love for animals.
- Grace often walked Rebecca on a leash, delighting the press and public.

Religious Persuasion

Coolidge's faith was private but influential, rooted in New England Protestant traditions.

- Raised in the Congregationalist Church, emphasizing simplicity and moral duty.
- Attended church regularly but avoided public displays of religiosity.
- His faith reinforced his belief in personal responsibility and limited government.
- Grace was also a devout Congregationalist, supporting church-related charities.

Interesting Anecdotes

Coolidge's life was filled with stories that highlighted his unique personality and era.

- Known as "Silent Cal," he once sat through a dinner with a woman betting she could make him talk; he responded minimally with, "You lose."
- During the Boston Police Strike, his firm stance earned him the nickname "Law and Order" Coolidge.
- Grace smuggled Rebecca the raccoon into the White House, hiding her in a bathtub to avoid staff disapproval.
- Coolidge napped daily in the White House, believing rest improved his decision-making clarity.

Ages at Death, Causes of Death, and Burial Locations

Calvin and Grace Coolidge lived relatively long lives after the presidency, passing away in their home state of Vermont.

- Calvin Coolidge: Died at 60 on January 5, 1933, of a heart attack in Northampton, Massachusetts. Buried in Plymouth Notch Cemetery, Plymouth, Vermont.
- Grace Coolidge: Died at 78 on July 8, 1957, of heart failure in Northampton, Massachusetts. Buried beside her husband in Plymouth Notch Cemetery.

Conclusion

Calvin Coolidge's presidency, marked by economic prosperity and minimalist governance, earned him the title of the "Silent Leader of Prosperity." His commitment to fiscal conservatism, tax cuts, and limited government fueled the Roaring Twenties but left vulnerabilities that surfaced in the 1929 crash. Grace Coolidge's warmth and advocacy complemented his reserved style, enhancing their public appeal. Together, they embodied a traditional, disciplined approach to leadership, leaving a legacy of stability and simplicity. While Coolidge's policies had mixed outcomes, his integrity and humility remain defining features of his place in American history.

Herbert Hoover: The Great Humanitarian

Introduction

Herbert Hoover, the 31st President of the United States, served from 1929 to 1933. Known as the "Great Humanitarian" for his relief efforts during and after World War I, Hoover's presidency was overshadowed by the onset of the Great Depression. A mining engineer and businessman turned public servant, he approached governance with a belief in individualism and limited government intervention, which shaped both his successes and challenges. This summary explores Hoover's life, presidency, and legacy, highlighting his contributions, personal traits, and the significant role of First Lady Lou Henry Hoover.

Early Life

Born on August 10, 1874, in West Branch, Iowa, Herbert Clark Hoover grew up in a modest Quaker community. His early life was marked by hardship and resilience, shaping his self-reliant worldview.

- Orphaned at age nine after both parents died (father, Jesse, in 1880; mother, Hulda, in 1884).
- Raised by relatives, primarily his uncle in Oregon.
- Worked odd jobs, including picking bugs off potato crops, to support himself.
- Developed an interest in engineering and geology, influenced by Oregon's rugged landscapes.

Family

Hoover's family life was rooted in his Quaker upbringing, which emphasized simplicity, service, and community. His marriage to Lou Henry was a partnership of shared intellect and adventure.

- Parents: Jesse Hoover, a blacksmith and farm equipment salesman, and Hulda Minthorn Hoover, a Quaker minister.
- Siblings: Older brother Theodore and younger sister Mary.
- Extended family: Lived with his uncle, John Minthorn, a physician and school superintendent, after being orphaned.

Children

Herbert and Lou Hoover had two sons, both of whom led accomplished lives influenced by their parents' values of service and education.

- Herbert Charles Hoover Jr. (1903–1969): Became a geophysical engineer and served as Under Secretary of State under Eisenhower.
- Allan Henry Hoover (1907–1993): Pursued a career in mining and finance, maintaining a low public profile.

Rise to Power

Hoover's ascent to the presidency was driven by his reputation as an efficient administrator and humanitarian, built through a successful career in mining and public service.

- Graduated from Stanford University in 1895 with a degree in geology.
- Worked as a mining engineer in Australia, China, and London, amassing wealth and global experience.
- Led humanitarian efforts during World War I, organizing food relief for millions in Europe as head of the Commission for Relief in Belgium.
- Served as U.S. Food Administrator (1917–1919) under President Wilson, managing domestic food conservation.
- Appointed Secretary of Commerce (1921–1928) under Presidents Harding and Coolidge, promoting economic modernization.

- Won the 1928 presidential election in a landslide against Democrat Al Smith.

Influences

Hoover's philosophy and policies were shaped by his Quaker roots, professional experiences, and belief in individual initiative.

- Quaker values of simplicity, charity, and community service guided his humanitarian efforts.
- Mining career instilled a problem-solving mindset and faith in technical expertise.
- Exposure to global crises (e.g., Boxer Rebellion in China, World War I) reinforced his commitment to efficient organization.
- Influenced by progressive ideas of efficiency and scientific management, but skeptical of centralized government power.

Party Affiliation

Hoover was a member of the Republican Party, aligning with its pro-business and limited-government principles during the 1920s.

- Joined the Republican Party in the early 20th century, reflecting his belief in individual enterprise.
- Supported progressive Republican policies, such as economic modernization, but opposed excessive government intervention.

- His affiliation shaped his presidency, emphasizing voluntary cooperation over federal mandates.

Presidency

Hoover's presidency (March 4, 1929–March 4, 1933) began with optimism but was defined by the Great Depression, which began with the stock market crash in October 1929. His response, rooted in limited government intervention, was widely criticized.

- Faced an unprecedented economic collapse, with unemployment reaching 25% by 1932.
- Promoted voluntary cooperation between businesses and local governments to stimulate recovery.
- Signed the Smoot-Hawley Tariff Act (1930), which raised tariffs and worsened global trade.
- Established the Reconstruction Finance Corporation (1932) to provide loans to banks and businesses.
- Opposed direct federal relief to individuals, believing it undermined self-reliance.
- Lost re-election in 1932 to Franklin D. Roosevelt in a landslide.

Accomplishments

Hoover's achievements spanned his pre-presidential and post-presidential careers, with significant contributions in humanitarianism and public administration.

- Organized relief for 10 million Belgians during World War I as head of the Commission for Relief in Belgium.

- Led U.S. food conservation efforts during World War I, saving millions from starvation.
- As Commerce Secretary, standardized industrial products, boosting efficiency.
- Established the Federal Farm Board (1929) to support agricultural prices.
- Initiated public works projects, including the Hoover Dam, to create jobs.
- Post-presidency, chaired the Hoover Commission (1947–1949), streamlining federal government operations.

First Lady's Contributions

Lou Henry Hoover, an educated and adventurous partner, redefined the role of First Lady with her activism and cultural contributions.

- Restored and furnished the White House with historical accuracy, emphasizing American heritage.
- Promoted the Girl Scouts, serving as national president (1922–1925, 1935–1937).
- Advocated for women's athletics and physical education.
- Supported charitable causes, including relief for unemployed families during the Depression.
- Translated a 16th-century mining text, *De Re Metallica*, with Herbert, earning scholarly recognition.

- Organized radio broadcasts to promote volunteerism and community service.

Personal Positive Traits

Hoover's strengths shaped his leadership style and humanitarian legacy.

- Exceptional organizational skills honed through engineering and relief work.
- Deep commitment to humanitarian causes, driven by Quaker values.
- Intellectual curiosity, evident in his global travels and scholarly pursuits.
- Resilience, overcoming early orphanhood to achieve professional success.

Personal Negative Traits

Hoover's weaknesses hindered his ability to address the Great Depression effectively.

- Stiff public demeanor, which limited his ability to inspire confidence.
- Overreliance on voluntarism and underestimating the need for federal intervention.
- Poor communication skills, failing to convey empathy during economic hardship.
- Stubborn adherence to principles, resisting policy shifts despite worsening conditions.

Effects of Traits on Presidency

Hoover's traits had a profound impact on his presidency, particularly during the Great Depression.

- His organizational skills led to innovative programs like the Reconstruction Finance Corporation, but their scope was limited.

- His reserved nature and reluctance to use media alienated the public, who sought reassurance.

- His commitment to limited government delayed direct relief, worsening public suffering, and damaging his reputation.

- His intellectual approach, while effective in technical roles, was less suited to political leadership in a crisis.

Pets

The Hoovers were animal lovers, and their White House was home to several pets, reflecting their warm family life.

- Dogs: Multiple breeds, including a German shepherd named King Tut and a fox terrier named Big Boy.

- Cats: Kept several, though less documented than their dogs.

- Other animals: Allan Hoover reportedly kept alligators in the White House as a teenager.

Religious Persuasion

Hoover's Quaker faith profoundly influenced his values and actions.

- Raised in the Religious Society of Friends (Quakers), emphasizing peace, simplicity, and service.
- Remained a practicing Quaker, though less overtly religious in public life.
- Faith guided his humanitarian work and belief in individual responsibility.

Interesting Anecdotes

Hoover's life was filled with unique stories that highlight his character and experiences.

- Survived the Boxer Rebellion in China (1900), organizing defenses for foreign residents in Tianjin.
- Once traveled 36 hours by camel to negotiate a mining deal in the Australian outback.
- As president, played "Hooverball," a volleyball-like game with a medicine ball, to stay fit.
- Secretly funded scholarships for students at Stanford, his alma mater, during the Depression.

Ages at Death

Herbert and Lou Hoover lived long lives, remaining active in public service after the presidency.

- Herbert Hoover: Died at age 90 on October 20, 1964.
- Lou Henry Hoover: Died at age 69 on January 7, 1944.

Causes of Death

Their causes of death reflected their advanced ages and health challenges.

- Herbert Hoover: Died of massive internal bleeding, likely from a gastrointestinal issue, in New York City.
- Lou Henry Hoover: Died of a heart attack in New York City.

Burial Locations

The Hoovers were laid to rest in places tied to their roots.

- Herbert Hoover: Buried at the Herbert Hoover Presidential Library and Museum in West Branch, Iowa.
- Lou Henry Hoover: Initially buried in Palo Alto, California; later reinterred with Herbert in West Branch, Iowa.

Conclusion

Herbert Hoover's legacy as the "Great Humanitarian" is a testament to his extraordinary contributions to global relief efforts and public administration, yet his presidency remains a cautionary tale of ideological rigidity in the face of economic crisis. His Quaker values, engineering mindset, and commitment to service shaped a remarkable career, from rescuing millions during World War I to modernizing American industry. However, his inability to adapt to the Great Depression's demands marred his presidency, leaving a mixed legacy. Lou Henry Hoover's trailblazing role as First Lady complemented his efforts, promoting education, heritage, and volunteerism. Together, their lives reflect a dedication to improving the world, even as they navigated personal and political challenges.

Franklin Delano Roosevelt: Architect of the New Deal and World War II Leadership

Description

Franklin Delano Roosevelt, commonly known as FDR, was the 32nd President of the United States, serving from 1933 to 1945. He is most remembered for leading the nation through the Great Depression with his New Deal policies and guiding the United States through much of World War II, establishing America as a global superpower. His unprecedented four-term presidency reshaped the role of the federal government and left a lasting legacy in American politics and society.

Introduction

Franklin Delano Roosevelt was born into privilege but became a champion of the common American, navigating the country through two of its greatest crises: the Great Depression and World War II. His leadership style, characterized by optimism, pragmatism, and bold experimentation, redefined the presidency. Despite physical challenges from polio, Roosevelt's resilience and communication skills, particularly through his "fireside chats," inspired a nation. His partnership with First Lady Eleanor Roosevelt amplified his administration's impact, as she became a transformative figure in her own right.

Early Life

Franklin Delano Roosevelt was born on January 30, 1882, in Hyde Park, New York, to a wealthy and prominent family. His early years were marked by privilege, education, and exposure to public service, shaping his worldview and ambitions.

- Grew up on the family estate, Springwood, in Hyde Park, with a sheltered but intellectually stimulating childhood.

- Educated by private tutors until age 14, then attended Groton School, an elite preparatory school in Massachusetts.

- Excelled at Harvard University (1900–1904), earning a bachelor's degree, and briefly attended Columbia Law School, passing the bar but not graduating.

- Developed an early interest in politics, inspired by his distant cousin, President Theodore Roosevelt.

Family

Roosevelt's family background and marriage played significant roles in his personal and political life.

- Son of James Roosevelt, a businessman, and Sara Ann Delano, a strong-willed matriarch who influenced Franklin's confidence.
- Married Anna Eleanor Roosevelt, his fifth cousin once removed, on March 17, 1905, in New York City.
- The marriage connected Franklin to Theodore Roosevelt, Eleanor's uncle, enhancing his political network.
- Sara Roosevelt's dominance created tensions in Franklin and Eleanor's marriage, though they maintained a strong partnership.

Children

Franklin and Eleanor Roosevelt had six children, five of whom survived to adulthood. Their children faced public scrutiny but also contributed to the family's political legacy.

- Anna Eleanor Roosevelt (1906–1975): Worked in journalism and public relations, later supporting her mother's activism.
- James Roosevelt (1907–1991): Served as a Marine in World War II and later as a congressman.

- Franklin Delano Roosevelt Jr. (1909): Died in infancy at eight months.

- Elliott Roosevelt (1910–1990): Served in the Army Air Forces during World War II and pursued a business career.

- Franklin Delano Roosevelt Jr. (1914–1988): Navy officer, congressman, and diplomat.

- John Aspinwall Roosevelt (1916–1981): Navy officer and businessman, less involved in politics.

Rise to Power

Roosevelt's political ascent was marked by strategic moves, personal resilience, and leveraging his family's name.

- Began political career in 1910 as a New York State Senator, advocating progressive reforms.

- Served as Assistant Secretary of the Navy (1913–1920) under President Woodrow Wilson, gaining national exposure.

- Ran as the Democratic vice-presidential candidate in 1920, though the ticket lost.

- Stricken with polio in 1921 at age 39, leaving him paralyzed from the waist down, his recovery efforts at Warm Springs, Georgia, showcased his determination.

- Elected Governor of New York (1929–1932), where his innovative relief programs foreshadowed the New Deal.

- Won the presidency in 1932, defeating Herbert Hoover in a landslide amid the Great Depression.

Influences

Roosevelt's policies and leadership were shaped by a blend of personal experiences, historical figures, and contemporary challenges.

- Inspired by Theodore Roosevelt's progressive activism and leadership style.
- Influenced by Woodrow Wilson's internationalism and vision for global cooperation.
- Shaped by the Progressive Era's emphasis on government intervention to address social and economic issues.
- Personal struggle with polio fostered empathy for the disadvantaged and a belief in collective action.
- Advised by a "Brain Trust" of intellectuals and economists who helped craft New Deal policies.

Party Affiliation

- A lifelong Democrat, Roosevelt transformed the Democratic Party into a coalition of urban workers, farmers, minorities, and intellectuals.
- Broke from conservative Democratic traditions by embracing federal intervention and progressive reforms.

- His New Deal coalition dominated American politics for decades, realigning the party toward economic and social justice.

Presidency

Roosevelt's presidency (1933–1945) was defined by bold responses to the Great Depression and World War II, fundamentally expanding the federal government's role.

- Took office during the Great Depression, with 25% unemployment and widespread bank failures.
- Launched the New Deal, a series of programs to provide relief, recovery, and reform.
- Declared a bank holiday in 1933 to restore confidence in the financial system.
- Delivered "fireside chats" via radio, directly addressing Americans to explain policies and inspire hope.
- Led the U.S. through World War II after the Pearl Harbor attack in 1941, mobilizing the economy and military.
- Played a key role in forming the Allied coalition, meeting with leaders like Winston Churchill and Joseph Stalin.
- Died in office on April 12, 1945, shortly before the end of World War II in Europe.

Accomplishments

- Established the New Deal, creating programs like Social Security, the Civilian Conservation Corps, and the Works Progress Administration.

- Restored confidence in the banking system through the Emergency Banking Act and Federal Deposit Insurance Corporation (FDIC).
- Signed the Wagner Act, protecting workers' rights to unionize and bargain collectively.
- Led the U.S. to victory in World War II, overseeing the production of war materials and the D-Day invasion.
- Co-founded the United Nations to promote international cooperation and peace.
- Enacted the Lend-Lease Act to support Allies before U.S. entry into World War II.
- Promoted the Good Neighbor Policy, improving U.S. relations with Latin America.

First Lady's Contributions

- Eleanor Roosevelt redefined the role of First Lady as a global advocate for human rights and social justice.
- Wrote a daily newspaper column, "My Day," sharing her views and connecting with the public.
- Championed civil rights, advocating for African Americans and other marginalized groups.
- Traveled extensively to support troops and boost morale during World War II.
- Played a key role in drafting the Universal Declaration of Human Rights at the United Nations.
- Promoted New Deal programs, particularly those aiding women, youth, and the poor.

- Served as a liaison between the public and the administration, relaying citizens' concerns to FDR.

Positive Traits and Effects

Roosevelt's personal strengths profoundly shaped his presidency.

- Optimism and charisma inspired hope, evident in his famous quote, "The only thing we have to fear is fear itself."
- Pragmatism allowed him to experiment with policies, adapting to economic and wartime challenges.
- Exceptional communication skills, particularly in fireside chats, built public trust and support.
- Resilience in overcoming polio demonstrated determination, endearing him to Americans facing hardship.
- These traits enabled him to rally the nation, pass sweeping legislation, and lead effectively in crisis.

Negative Traits and Effects

Roosevelt's flaws occasionally hindered his presidency.

- Tendency to centralize power led to criticism of overreach, notably in the 1937 court-packing plan.
- Reluctance to confront Southern Democrats limited progress on civil rights, frustrating activists.
- Secretiveness about his health concealed the extent of his physical decline, raising questions about transparency.

- Overconfidence sometimes led to missteps, such as underestimating opposition to New Deal reforms.
- These traits sparked controversies, including Supreme Court battles and tensions within his coalition.

Pets

The Roosevelts were animal lovers, and their pets were a visible part of their White House life.

- Fala, a Scottish Terrier, was FDR's most famous pet, often accompanying him and appearing in the media.
- Other pets included a German Shepherd named Major and various dogs owned by the family.
- Fala became a cultural icon, symbolizing Roosevelt's relatable, down-to-earth persona.

Religious Persuasion

- Roosevelt was raised in the Episcopal Church and remained a practicing Episcopalian throughout his life.
- His faith influenced his moral outlook, emphasizing compassion and social responsibility.
- Eleanor shared his Episcopalian background but focused more on universal ethical principles in her activism.
- FDR's religious beliefs were private, but informed his public calls for hope and unity.

Interesting Anecdotes

- FDR's polio diagnosis in 1921 was initially kept private; he used a wheelchair or braces but was rarely photographed this way to maintain a strong public image.

- His dog Fala once sparked a political controversy when Republicans falsely claimed FDR sent a Navy destroyer to retrieve him, which FDR humorously debunked in a campaign speech.

- Roosevelt loved sailing and collecting naval memorabilia, reflecting his lifelong passion for the sea.

- He designed his own wheelchair, a modified kitchen chair with wheels, to maintain independence.

- Eleanor and Franklin's marriage survived his 1918 affair with Lucy Mercer, evolving into a political partnership.

Ages at Death, Causes of Death, and Burial Locations

- **Franklin Delano Roosevelt**: Died at age 63 on April 12, 1945, from a cerebral hemorrhage in Warm Springs, Georgia. Buried at the family estate, Springwood, in Hyde Park, New York.

- **Eleanor Roosevelt**: Died at age 78 on November 7, 1962, from aplastic anemia and tuberculosis in New York City. Buried beside Franklin at Springwood, Hyde Park, New York.

Conclusion

Franklin Delano Roosevelt's presidency was a transformative era in American history, marked by innovative domestic policies and decisive wartime leadership. His New Deal alleviated suffering and redefined the government's role, while his global vision shaped the post-war world. Eleanor Roosevelt's groundbreaking activism complemented his achievements, advancing human rights and social justice. Despite personal and political flaws, FDR's optimism, resilience, and adaptability made him a towering figure. Together, Franklin and Eleanor left an enduring legacy of compassion, courage, and progress that continues to inspire.

Harry S. Truman: The Man from Missouri

Description

Harry S. Truman, the 33rd President of the United States, was a plain-spoken, decisive leader who rose from humble beginnings to guide the nation through the end of World War II and the early Cold War. Known for his integrity and "the buck stops here" philosophy, Truman's presidency was marked by bold decisions that shaped the modern world.

Introduction

Harry S. Truman served as president from April 12, 1945, to January 20, 1953, assuming office after Franklin D. Roosevelt's death. His tenure included pivotal moments like the atomic bomb, the founding of the United Nations, and the beginning of the Cold War. A farmer, soldier, and small-town politician, Truman's unpretentious demeanor belied his ability to make tough choices in turbulent times.

Early Life

Born on May 8, 1884, in Lamar, Missouri, Truman grew up in a modest farming family. His early years were shaped by rural life and a strong work ethic.

- Moved to Independence, Missouri, at age six, where he spent most of his childhood.
- Struggled with poor eyesight, requiring thick glasses from a young age.
- Worked on the family farm, developing a practical, no-nonsense outlook.
- Did not attend college due to financial constraints, instead pursuing various jobs, including bank clerk and railroad timekeeper.

Family

Truman's family life was anchored by close relationships and a deep commitment to his wife and daughter.

- Parents: John Anderson Truman and Martha Ellen Young Truman, both of modest means.
- Siblings: Two younger siblings, John Vivian and Mary Jane Truman.
- Married Bess Wallace in 1919 after a long courtship that began in childhood.
- Maintained a close relationship with his mother, who lived with the Trumans during his presidency.

Children

Truman had one child, a daughter, who became a significant part of his personal and public life.

- Margaret Truman, born February 17, 1924, was the couple's only child.

- Margaret pursued a career as a singer and later a writer, publishing biographies and mystery novels.

- Truman was a devoted father, often writing letters to Margaret during his presidency, offering advice and support.

Rise to Power

Truman's political career began locally and grew through perseverance and connections.

- Served as a captain in World War I, gaining leadership experience in artillery.

- Elected as a Jackson County judge (an administrative role) in 1922, serving two terms.

- Backed by Kansas City's Pendergast political machine, which helped him win a U.S. Senate seat in 1934.

- Gained national prominence chairing the Truman Committee, investigating wartime spending waste.

- Selected as Franklin D. Roosevelt's vice-presidential running mate in 1944, becoming vice president in January 1945.

- Assumed the presidency on April 12, 1945, after Roosevelt's sudden death.

Influences

Truman's worldview was shaped by personal experiences and historical figures.

- Admired Abraham Lincoln for his honesty and leadership during a crisis.
- Influenced by his Baptist upbringing, he emphasizes duty and community.
- World War I service instilled a sense of discipline and camaraderie.
- Studied history independently, drawing inspiration from leaders like Andrew Jackson for their decisiveness.

Party Affiliation

Truman was a lifelong Democrat, aligned with the party's progressive and populist wings.

- Supported New Deal policies but favored fiscal responsibility.
- Championed working-class issues, reflecting his rural Missouri roots.
- Navigated tensions between liberal and conservative Democrats during his presidency.

Presidency

Truman's presidency (1945–1953) was defined by the end of World War II and the onset of the Cold War, requiring bold and often controversial decisions.

- Authorized the use of atomic bombs on Hiroshima and Nagasaki in August 1945 to end World War II.

- Oversaw the transition to a peacetime economy, addressing labor strikes and inflation.

- Faced domestic challenges, including civil rights tensions and anti-communist fears led by Senator Joseph McCarthy.

- Dealt with the Korean War (1950–1953), a major Cold War conflict, and controversially fired General Douglas MacArthur for insubordination.

- Left office with low approval ratings but later gained recognition for his principled leadership.

Accomplishments

Truman's key achievements had lasting impacts on domestic and foreign policy.

- Established the Truman Doctrine (1947), providing aid to nations resisting communism, shaping Cold War policy.

- Implemented the Marshall Plan (1948), rebuilding Western Europe and preventing Soviet expansion.

- Desegregated the U.S. military (1948), a major step toward civil rights.

- Recognized the state of Israel in 1948, supporting its creation.

- Oversaw the formation of NATO (1949), strengthening Western alliances.

- Enacted the Fair Deal, expanding social programs like housing and education, though Congress blocked much of it.

First Lady's Contributions

Bess Truman, the First Lady, preferred a private role but made subtle contributions.

- Hosted social events at the White House, maintaining traditions despite her dislike for publicity.
- Supported charitable causes quietly, including children's hospitals and veterans' organizations.
- Influenced Truman's decisions through private counsel, acting as a trusted confidante.
- Oversaw White House renovations, ensuring the preservation of its historical integrity.

Positive Traits

Truman's personal strengths shaped his leadership style.

- Decisiveness: Made tough calls, like using the atomic bomb, with clarity.
- Integrity: Known for honesty, reflected in his "buck stops here" motto.
- Work ethic: Tackled challenges tirelessly, from farm work to presidency.
- Loyalty: Stood by aides and allies, fostering trust within his administration.

Negative Traits

Truman's flaws sometimes complicated his presidency.

- Temper: His blunt, occasionally abrasive style alienated some colleagues and the press.
- Loyalty to a fault: Defended allies like the Pendergast machine, raising ethical questions.
- Inexperience: Lacked formal education and initially struggled with the presidency's scope.
- Stubbornness: Clung to decisions, like MacArthur's firing, despite public backlash.

Effects of Traits on Presidency

Truman's traits had a mixed impact on his leadership.

- His decisiveness enabled swift action in crises like the Korean War but led to controversial choices, such as the atomic bomb.
- Integrity earned him respect over time, though his temper strained relations with Congress and the media.
- Inexperience initially caused missteps, but his work ethic helped him adapt to global challenges.
- Stubbornness solidified policies like the Truman Doctrine but fueled domestic criticism, lowering his approval ratings.

Pets

The Trumans had few pets, reflecting their practical lifestyle.

- Margaret owned a dog named Feller, a cocker spaniel gifted to the family in 1947.
- Feller was sent to a farm after media attention, as the Trumans preferred privacy.
- No other significant pets were documented during their White House years.

Religious Persuasion

Truman's faith influenced his moral framework.

- Raised as a Southern Baptist, attending church regularly in Missouri.
- Believed in Christian values like duty and charity but rarely discussed religion publicly.
- Supported religious freedom, reflected in his recognition of Israel and ecumenical policies.

Interesting Anecdotes

Truman's life was filled with colorful moments that revealed his character.

- As a young man, he proposed to Bess via letter, only to be initially rejected; he persisted for years.
- Played the piano enthusiastically, once performing at a White House event for dignitaries.
- Survived an assassination attempt in 1950 by Puerto Rican nationalists at Blair House, showing calm under pressure.

- Took daily walks in Washington, D.C., often at a brisk 120 steps per minute, surprising reporters.

Ages at Death

- Harry S. Truman died at age 88 on December 26, 1972.
- Bess Truman died at age 97 on October 18, 1982.

Causes of Death

- Harry S. Truman: Died of heart failure and old age, complicated by a minor lung infection.
- Bess Truman: Died of congestive heart failure.

Burial Locations

- Both Harry and Bess Truman are buried at the Harry S. Truman Presidential Library and Museum in Independence, Missouri.
- Their graves are located in the library's courtyard, a site chosen to reflect their connection to Missouri.

Conclusion

Harry S. Truman's legacy as "The Man from Missouri" rests on his ability to lead during a transformative era. From ending World War II to laying the foundations of the Cold War, his presidency shaped global and domestic policy. His straightforward, principled approach, coupled with Bess's quiet support, defined their time in the White House. Though initially unpopular, Truman's decisions, like the Marshall Plan and military desegregation, earned him posthumous admiration as a leader who faced unprecedented challenges with courage and conviction.

Dwight D. Eisenhower: Supreme Allied Commander and Cold War Leader

Description

Dwight David Eisenhower, the 34th President of the United States, was a military hero and statesman who led the nation through the early Cold War. Known for his strategic leadership as Supreme Allied Commander during World War II, Eisenhower brought a steady, pragmatic approach to governance, emphasizing peace, prosperity, and global stability. His presidency focused on containing communism, fostering economic growth, and advancing infrastructure, leaving a lasting legacy in American and global history.

Introduction

Dwight D. Eisenhower, often called "Ike," served as president from 1953 to 1961, guiding the United States through a transformative period marked by post-war recovery and Cold War tensions. Before his presidency, he was a five-star general who orchestrated the D-Day invasion, earning international acclaim. His leadership style, rooted in discipline and collaboration, shaped his domestic and foreign policies, making him a pivotal figure in 20th-century America.

Early Life

Dwight D. Eisenhower's early life was marked by modest beginnings, strong family values, and experiences that shaped his character.

- Born on October 14, 1890, in Denison, Texas, Eisenhower grew up in Abilene, Kansas, in a modest, religious household.
- Third of seven sons.
- Developed a strong work ethic and a love for sports and history.
- Early life in a tight-knit community instilled values of duty and resilience.
- These values guided his military and political career.

Family

Eisenhower's upbringing was deeply influenced by his family's faith, values, and modest lifestyle.

- He was raised in a large family with strong ties to the River Brethren, a pacifist Christian sect.
- His parents, David Jacob Eisenhower and Ida Elizabeth Stover, emphasized frugality and faith in their household.

- His mother's pacifism contrasted with his eventual military path, but her influence fostered his moral grounding.
- The family's modest circumstances shaped his practical and unpretentious outlook on life.

Children

Eisenhower's marriage and family life were marked by both profound joy and deep personal loss.

- He married Mamie Geneva Doud in 1916.
- The couple had two sons during their marriage.
- Their first son, Doud Dwight ("Ikky"), died at the age of three from scarlet fever, a tragedy that deeply affected them both.
- Their second son, John Sheldon Doud Eisenhower, was born in 1922.
- John became a military officer, historian, and diplomat, carrying on his father's legacy of public service.

Rise to Power

Eisenhower's rise to national and international prominence was driven by his military achievements and leadership during pivotal moments in history.

- He graduated from West Point in 1915, marking the start of his military career.
- He served under Generals John J. Pershing and Douglas MacArthur, honing his strategic and leadership skills.

- During World War II, he became Supreme Allied Commander and led the D-Day invasion, a role that catapulted him to global fame.

- After the war, he was courted by both major U.S. political parties to run for president.

- He ultimately chose the Republican Party and won the 1952 presidential election, capitalizing on his reputation as a war hero.

Influences

Eisenhower's worldview and leadership style were shaped by his faith, his mentors, and the political realities of his time.

- He was strongly influenced by his religious upbringing, which instilled in him a deep moral compass.

- His mother's faith played a significant role in guiding his values and ethical outlook.

- Military mentors such as Generals John J. Pershing and George C. Marshall taught him strategic discipline and leadership.

- The geopolitical tensions of the Cold War, including the threat of nuclear conflict, shaped his priorities as a leader.

- He emphasized diplomacy and deterrence, seeking to balance military strength with sustained peace efforts.

Party Affiliation

Eisenhower's political path reflected his pragmatic approach and broad appeal to the American electorate.

- He was a Republican, though his political alignment was pragmatic rather than strictly ideological.
- Initially apolitical, he declined overtures from the Democratic Party before entering politics.
- He decided to run as a Republican in the 1952 presidential election.
- His moderate stance attracted a wide range of voters from across the political spectrum.
- He prioritized fiscal conservatism and international cooperation over rigid partisan positions.

Presidency

Eisenhower's presidency spanned a period of Cold War tension, domestic growth, and gradual social change.

- He served two terms as president from January 20, 1953, to January 20, 1961.
- His administration navigated the complexities of the Cold War while maintaining domestic prosperity.
- He prioritized containing communism and strengthening the North Atlantic Treaty Organization (NATO).
- He worked to promote economic growth across the United States.

- He approached civil rights issues with caution, balancing progress with political realities.
- His presidency is remembered for major infrastructure development, including the interstate highway system, and for practicing diplomatic restraint.

Accomplishments

- Orchestrated the Korean War armistice in 1953, ending active conflict.
- Authorized the Interstate Highway System, revolutionizing American transportation.
- Strengthened NATO and established the Eisenhower Doctrine to counter Soviet influence in the Middle East.
- Signed the Civil Rights Act of 1957, the first such legislation since Reconstruction.
- Created NASA in 1958, advancing U.S. space exploration.
- Maintained economic stability, with low unemployment and inflation during the 1950s boom.
- Promoted "Atoms for Peace," advocating peaceful nuclear energy use.

First Lady's Contributions

- Mamie Eisenhower hosted White House social events, restoring elegance post-war with her charm and style.
- Championed military families, drawing on her experience as an Army wife.

- Supported the American Heart Association, raising awareness for heart disease.
- Promoted inclusivity by inviting diverse groups to White House functions.
- Maintained a warm public image, boosting national morale during tense times.

Positive Traits

Eisenhower's leadership style combined composure, foresight, and teamwork, enabling him to guide the nation through challenging times.

- He maintained a calm demeanor that inspired trust among allies, advisors, and the public.
- He demonstrated strategic foresight in both domestic policy and foreign affairs.
- He practiced collaborative leadership, building consensus to strengthen his administration's effectiveness.
- He delegated responsibilities effectively, empowering capable advisors to carry out his vision.
- His military background shaped a disciplined and organized approach to governance.
- These qualities allowed him to navigate Cold War crises and maintain national unity during turbulent periods.

Negative Traits

Eisenhower's reserved approach to contentious issues sometimes slowed social progress and affected perceptions of his leadership.

- He took a cautious approach to civil rights, which frustrated activists seeking faster change.
- He avoided bold action against segregation during his presidency.
- His reluctance to confront Senator Joseph McCarthy early allowed McCarthyism to flourish temporarily.
- His reserved personality occasionally limited his public engagement, making him seem distant.
- These traits contributed to slower progress on key social issues.
- They also, at times, undermined the perceived dynamism of his leadership.

Pets

Eisenhower's quiet affection for animals was evident in his relationship with the family dog, Heidi.

- The Eisenhowers owned a Weimaraner named Heidi, gifted to them by a breeder in 1955.
- Heidi roamed freely around the White House grounds during Eisenhower's presidency.
- She was a beloved companion to the Eisenhower family.

- Eisenhower's fondness for animals was understated but deeply genuine.
- Heidi provided him with a private source of comfort amid the pressures of leadership.

Religious Persuasion

Eisenhower's religious beliefs evolved over time, but faith remained a guiding influence in his life and leadership.

- He was raised in the River Brethren, a pacifist Christian denomination.
- After his marriage, he became a Presbyterian.
- His faith emphasized duty, morality, and personal responsibility.
- Religious values played a role in shaping his leadership style and decisions.
- He promoted national prayer and supported adding the phrase "under God" to the Pledge of Allegiance.
- He viewed religion as a unifying force but avoided overtly dogmatic displays.

Interesting Anecdotes

Eisenhower enjoyed hobbies and personal traditions that offered relaxation and connection beyond politics.

- He was an avid golfer and even had a putting green installed at the White House.
- He once painted a portrait of his wife, Mamie, showcasing a lesser-known artistic side.

- During World War II, he carried a lucky coin, a habit he continued throughout his presidency.
- He enjoyed grilling steaks at Camp David during downtime.
- He often hosted informal gatherings with world leaders at Camp David, blending diplomacy with hospitality.

Ages at Death

Eisenhower and his wife, Mamie, lived long lives marked by public service and personal devotion.

- Dwight D. Eisenhower died at the age of 78 on March 28, 1969.
- Mamie Eisenhower died at the age of 82 on November 1, 1979.
- Their passing marked the end of an era shaped by their leadership, resilience, and dedication to the nation.

Causes of Death

Eisenhower and Mamie's final years were marked by declining health, which ultimately led to their passing.

- Eisenhower died from congestive heart failure following a series of heart attacks.
- His health had been weakened over the years by chronic medical issues.
- Mamie Eisenhower passed away from a stroke in her later years.
- Her death followed a period of declining health.

Burial Locations

Eisenhower and Mamie were laid to rest together in a place that honors both their legacy and their roots.

- Both Dwight and Mamie Eisenhower are buried at the Dwight D. Eisenhower Presidential Library and Museum in Abilene, Kansas.
- Their final resting place is near Eisenhower's childhood home.
- The site includes a chapel where they rest side by side.
- Their burial together symbolizes their lifelong partnership and shared dedication to the nation.

Conclusion

Dwight D. Eisenhower's legacy as a military leader and president endures through his contributions to global stability and American infrastructure. His steady hand guided the nation through the Cold War's early years, while Mamie's warmth complemented his public image. Despite criticisms of his cautious approach to social issues, Eisenhower's pragmatic leadership and strategic vision cemented his place as a transformative figure, remembered for his role as a wartime commander and a president who prioritized peace and prosperity.

John F. Kennedy: The Camelot President

Introduction

John F. Kennedy, the 35th President of the United States, is most remembered for his charismatic leadership and vision of a progressive America, often referred to as the "Camelot" era due to its idealized portrayal. His presidency, though brief, was marked by significant events like the Cuban Missile Crisis and the push for civil rights, leaving a lasting legacy.

Description

John Fitzgerald Kennedy, commonly known as JFK, was a charismatic, youthful leader whose eloquence and idealism captivated the nation. His presidency focused on global diplomacy, civil rights, and space exploration, but it was tragically cut short by his assassination.

Early Life

John F. Kennedy's early life was shaped by privilege, education, and persistent health challenges.

- He was born on May 29, 1917, in Brookline, Massachusetts.
- He grew up in a wealthy, politically connected family.
- He faced lifelong health challenges, including Addison's disease, which he managed throughout his life.
- He attended elite schools, including Choate and Harvard University.
- He graduated from Harvard in 1940 with a degree in international affairs.

Family

Kennedy's upbringing was deeply influenced by his large, accomplished family and their strong Irish-American Catholic heritage.

- He came from a prominent Irish-American Catholic family.
- His father, Joseph P. Kennedy Sr., was a businessman and former U.S. ambassador.
- His mother, Rose Fitzgerald Kennedy, was a philanthropist.
- He had eight siblings, creating a lively and competitive household.
- His brother Robert F. Kennedy later served as U.S. Attorney General.
- His brother Edward "Ted" Kennedy became a long-serving U.S. senator.

Children

John F. Kennedy and his wife, Jacqueline "Jackie" Kennedy, had four children, though only two survived to adulthood:

- Caroline Bouvier Kennedy, born November 27, 1957, later became an author and diplomat.
- John F. Kennedy Jr., born November 25, 1960, became a lawyer and publisher before his death in 1999.
- Patrick Bouvier Kennedy, born August 7, 1963, died two days later.
- An unnamed daughter, stillborn in 1956.

Rise to Power

Kennedy's rise in politics was marked by rapid success, public appeal, and a flair for communication.

- His political career began in 1946 when he was elected to the U.S. House of Representatives from Massachusetts.
- He served three terms in the House before winning a U.S. Senate seat in 1952.
- His national profile grew due to his charisma and skillful use of the media.
- He authored *Profiles in Courage*, which won a Pulitzer Prize.
- In 1960, he won the presidency by narrowly defeating Richard Nixon in a close election.

Influences

Kennedy's worldview was shaped by his upbringing, military service, and exposure to global politics.

- He was influenced by his father's ambition, which drove his own political aspirations.
- His Catholic faith provided a moral framework that guided his decisions.
- The geopolitical tensions of the Cold War sharpened his focus on international affairs.
- His World War II service, where he earned a Navy and Marine Corps Medal for heroism, shaped his views on leadership and sacrifice.
- Influential thinkers like historian Arthur Schlesinger Jr. and interactions with global leaders informed his progressive policy vision.

Party Affiliation

Kennedy's political allegiance reflected a lifelong commitment to Democratic values.

- He was a lifelong member of the Democratic Party.
- He aligned himself with the party's liberal wing, supporting progressive reforms.
- His policies championed social reforms, civil rights, and economic growth.
- He navigated internal party tensions between progressive and conservative factions.

Presidency

Kennedy's presidency blended idealism with the realities of Cold War politics.

- He served as president from 1961 to 1963, with a term defined by global and domestic challenges.
- He faced Cold War crises, including the Bay of Pigs invasion and the Cuban Missile Crisis.
- His administration promoted major initiatives such as the Peace Corps and the space race.
- His "New Frontier" agenda sought advancements in education, healthcare, and equality.
- He pushed for civil rights while balancing political opposition and public opinion.

Accomplishments

- Established the Peace Corps in 1961 to promote global volunteerism.
- Successfully navigated the Cuban Missile Crisis in 1962, averting nuclear war.
- Signed the Nuclear Test Ban Treaty in 1963 to limit nuclear proliferation.
- Advanced the U.S. space program, setting the goal of landing a man on the moon.
- Supported civil rights, proposing legislation that led to the 1964 Civil Rights Act.

- Launched the Alliance for Progress to foster economic development in Latin America.

First Lady's Contributions

- Restored the White House, making it a showcase of American history and culture.
- Promoted the arts, hosting cultural events and supporting historic preservation.
- Enhanced U.S. diplomacy through her charm and multilingual skills during international trips.
- Created the White House Historical Association to preserve the mansion's legacy.
- Championed the preservation of Lafayette Square and other historic sites.

Positive Traits

Kennedy's leadership was fueled by qualities that inspired trust, unity, and ambition.

- His charisma, wit, and optimism motivated a generation and fostered a sense of national purpose.
- He communicated effectively through television and speeches, strengthening his connection with the public.
- His intellectual curiosity encouraged him to seek diverse perspectives from advisors.
- His openness to new ideas led to innovative policies, particularly in foreign affairs.

Negative Traits

Despite his strengths, Kennedy faced personal and professional challenges that sometimes undermined his effectiveness.

- His impulsiveness occasionally led to mistakes, most notably the Bay of Pigs invasion.
- He managed private health struggles, which were concealed from the public.
- Rumored personal indiscretions, though not widely known at the time, had the potential to damage his reputation.
- His cautious approach to civil rights frustrated activists and delayed immediate reforms.

Effects of Traits on Presidency

Kennedy's personality traits had a direct impact on the successes and shortcomings of his presidency.

- His charisma helped rally public support for ambitious initiatives like the moon landing.
- His impulsiveness contributed to early foreign policy missteps.
- His health issues, kept secret, required careful image management to maintain a perception of vitality.
- His cautious stance on civil rights balanced political pressures but slowed progress on equality.

Pets

The Kennedy family had several pets, reflecting their lively household:

- Dogs: Charlie (Welsh terrier), Pushinka (a gift from Soviet Premier Khrushchev), and others.
- Ponies: Macaroni and Tex, often ridden by Caroline and John Jr.
- Other animals: A canary (Robin), hamsters, and a rabbit named Zsa Zsa.

Religious Persuasion

Kennedy was a practicing Roman Catholic, the first Catholic U.S. president. His faith influenced his moral outlook, though he emphasized the separation of church and state to counter anti-Catholic prejudice during his campaign.

Interesting Anecdotes

- During the Cuban Missile Crisis, Kennedy secretly recorded White House meetings, providing historians with rare insights into his decision-making.
- As a young naval officer, he swam for hours to save his crew after their PT-109 boat was sunk in 1943.
- Jackie Kennedy's televised White House tour in 1962 drew millions, showcasing her cultural influence.
- Kennedy's quip, "I'm the man who accompanied Jacqueline Kennedy to Paris," highlighted her diplomatic charm.

Ages at Death

- John F. Kennedy died at age 46 on November 22, 1963.
- Jacqueline Kennedy Onassis died at age 64 on May 19, 1994.

Causes of Death

- John F. Kennedy was assassinated by a gunshot in Dallas, Texas, by Lee Harvey Oswald (though conspiracy theories persist).
- Jacqueline Kennedy Onassis died of non-Hodgkin's lymphoma in New York City.

Burial Locations

- John F. Kennedy was buried at Arlington National Cemetery, Virginia, with an eternal flame marking his grave.
- Jacqueline Kennedy Onassis was buried beside him at Arlington National Cemetery.

Conclusion

John F. Kennedy's life and presidency, though brief, left an indelible mark on American history. His vision of progress, eloquence, and leadership during crises like the Cuban Missile Crisis defined his legacy as the "Camelot President." Alongside Jacqueline's cultural and diplomatic contributions, their partnership shaped a transformative era. Despite personal and political challenges, Kennedy's idealism continues to inspire, cemented by his tragic death and enduring influence.

Lyndon B. Johnson: Architect of the Great Society

Description

Lyndon Baines Johnson, the 36th President of the United States, was a towering figure in American politics, known for his ambitious domestic agenda and complex legacy. A Texan with a larger-than-life personality, he rose from humble beginnings to become a master legislator and president, shaping modern America through landmark civil rights and social welfare legislation. His presidency, however, was also marred by the Vietnam War, which overshadowed many of his achievements.

Introduction

Lyndon B. Johnson's life is a story of determination, political skill, and transformative leadership. Born into rural poverty, he navigated the rough terrain of Texas politics to reach the White House. His presidency (1963–1969) was defined by the "Great Society," a sweeping set of programs aimed at eliminating poverty and racial injustice. Yet, his escalation of the Vietnam War divided the nation and clouded his legacy. Johnson's life reflects both the promise and the challenges of mid-20th-century America.

Early Life

Lyndon B. Johnson was born on August 27, 1908, in Stonewall, Texas, in a small farmhouse near the Pedernales River. Growing up in the rural Hill Country, he experienced financial hardship and the realities of rural life, which shaped his empathy for the disadvantaged.

- Attended local schools, often struggling academically but excelling in debate and leadership.

- Graduated from Johnson City High School in 1924.

- Worked odd jobs, including as a laborer, before enrolling at Southwest Texas State Teachers College (now Texas State University), earning a teaching degree in 1930.

- Taught briefly at a segregated Mexican-American school, an experience that influenced his later commitment to civil rights.

Family

Johnson came from a family with deep Texas roots, steeped in populist and Democratic traditions.

- Father: Samuel Ealy Johnson Jr., a farmer and state legislator who struggled financially but instilled political ambition in Lyndon.

- Mother: Rebekah Baines Johnson, a cultured woman who encouraged education and ambition.

- Siblings: Four siblings—Rebekah, Josefa, Sam Houston, and Lucia—grew up in a tight-knit but financially strained household.
- Johnson's family background fostered his drive to overcome poverty and achieve political success.

Children

Lyndon married Claudia Alta "Lady Bird" Taylor in 1934, and they had two daughters.

- Lynda Bird Johnson (born 1944): Married Charles S. Robb, later a Virginia governor and senator; active in public life.
- Luci Baines Johnson (born 1947): Married Patrick Nugent (later divorced), then Ian Turpin; involved in family business and philanthropy.
- Both daughters were active in supporting their father's political career and their mother's initiatives, though they maintained relatively private lives compared to their parents.

Rise to Power

Johnson's ascent was marked by relentless ambition, political savvy, and mentorship from key figures.

- Began as a congressional aide to Rep. Richard Kleberg in 1931, learning the intricacies of Washington politics.
- Appointed Texas director of the National Youth Administration (1935–1937), gaining statewide recognition.

- Elected to the U.S. House of Representatives in 1937, representing Texas's 10th district.

- Won a U.S. Senate seat in 1948 after a controversial election, earning the nickname "Landslide Lyndon" due to a narrow, disputed victory.

- Became Senate Majority Leader in 1955, mastering legislative strategy and coalition-building.

- Ran unsuccessfully for the Democratic presidential nomination in 1960 but accepted the vice-presidential slot under John F. Kennedy.

- Assumed the presidency on November 22, 1963, after Kennedy's assassination.

Influences

Johnson's worldview was shaped by personal experiences and political mentors.

- His father's populist leanings and struggles with poverty instilled a commitment to helping the disadvantaged.

- Teaching Mexican-American students exposed him to systemic inequalities, influencing his civil rights advocacy.

- Mentors like House Speaker Sam Rayburn and President Franklin D. Roosevelt inspired his legislative approach and belief in government's role in social progress.

- The New Deal era profoundly influenced his vision for the Great Society, emphasizing federal intervention to address poverty and inequality.

Party Affiliation

- Lifelong Democrat, deeply rooted in the party's Southern and progressive wings.
- Aligned with the New Deal liberalism of Franklin D. Roosevelt, advocating for government programs to aid the poor.
- Navigated tensions between Southern conservative Democrats and the party's growing liberal faction, particularly on civil rights.
- His leadership helped transform the Democratic Party into a champion of civil rights, though it alienated some Southern supporters.

Presidency

Johnson's presidency (1963–1969) was a pivotal era, marked by bold domestic reforms and the escalating Vietnam War. Sworn in aboard Air Force One hours after Kennedy's assassination, he vowed to continue Kennedy's agenda but quickly made it his own.

- Pushed through landmark legislation, including the Civil Rights Act of 1964 and the Voting Rights Act of 1965, dismantling legal segregation.
- Launched the Great Society, a series of programs addressing poverty, education, healthcare, and urban development.
- Escalated U.S. involvement in Vietnam, increasing troop levels from 16,000 in 1963 to over 500,000 by 1968, leading to widespread protests and division.

- Faced challenges like urban riots, anti-war movements, and economic strain from funding both the war and domestic programs.
  - Declined to run for re-election in 1968, citing national divisions, and retired to his Texas ranch.

Accomplishments

- Passed the Civil Rights Act of 1964, banning discrimination in public accommodations and employment.
- Signed the Voting Rights Act of 1965, ensuring African Americans' right to vote.
- Established Medicare and Medicaid, providing healthcare for the elderly and low-income Americans.
- Launched the War on Poverty, creating programs like Head Start and Job Corps.
- Passed the Elementary and Secondary Education Act of 1965, increasing federal funding for schools.
- Signed the Immigration Act of 1965, abolishing restrictive quotas and reshaping U.S. demographics.
- Created the Department of Housing and Urban Development and the Department of Transportation.
- Advanced environmental protection through the Clean Air Act and Water Quality Act.

First Lady's Contributions

- Lady Bird Johnson championed environmental conservation and urban beautification.

- Promoted the Highway Beautification Act of 1965, reducing billboards and enhancing roadside aesthetics.
- Supported Head Start, advocating for early childhood education.
- Planted millions of flowers in Washington, D.C., earning the nickname "Lady Bird" for her nature advocacy.
- Founded the First Lady's Committee for a More Beautiful Capital, improving urban spaces.
- Recorded a detailed White House diary, providing a historical record of the Johnson presidency.
- Established the Lady Bird Johnson Wildflower Center, promoting native plant conservation.

Positive Traits and Effects

Johnson's strengths were instrumental in his legislative success but also shaped his presidency's challenges.

- Masterful negotiator and persuader, known for the "Johnson Treatment"—intense, personal lobbying that secured legislative victories.
- Deep empathy for the poor and marginalized, rooted in his early life, drove his commitment to civil rights and social programs.
- Tireless work ethic and political instincts allowed him to navigate a fractious Congress and pass transformative laws.
- Visionary leadership created a lasting framework for social welfare and equality, reshaping American society.

- These traits enabled historic achievements but sometimes led to overconfidence in managing complex issues like Vietnam.

Negative Traits and Effects

Johnson's flaws contributed to the controversies of his presidency.

- Domineering personality could alienate allies and create resentment among colleagues.

- Tendency toward secrecy and mistrust, particularly regarding Vietnam, eroded public confidence.

- Stubborn commitment to the Vietnam War, driven by fear of appearing weak, deepened national divisions and overshadowed domestic successes.

- Impulsiveness in decision-making sometimes led to poorly considered policies or escalation of conflicts.

- These traits fueled anti-war sentiment and limited his ability to unify the nation during turbulent times.

Pets

The Johnsons were fond of animals, and their pets were a visible part of White House life.

- Beagles: Him and Her, the most famous White House pets, often featured in media.

- Beagle puppies: Freckles and Edgar, offspring of Him and Her.

- Blanco: A white collie gifted to the Johnsons.

- Yuki: A mixed-breed dog adopted by Luci Baines Johnson, often seen with LBJ.
- The pets humanized the Johnsons but drew criticism when LBJ was photographed lifting Him by the ears, sparking animal cruelty concerns.

Religious Persuasion

- Raised in a Baptist household, but joined the Disciples of Christ, his wife's denomination, after marriage.
- Attended services irregularly but valued faith as a moral guide, often referencing it in speeches.
- His religious background influenced his moral commitment to social justice and equality, though he was not overtly devout.
- Lady Bird was more active in her faith, attending services regularly and drawing on it for personal strength.

Interesting Anecdotes

- Johnson was notorious for his colorful language and blunt humor, once joking about a political opponent's lack of intelligence in vivid terms.
- During a 1964 campaign event, he drove reporters in his car at high speed around his Texas ranch, sipping a drink and steering with one hand, showcasing his bold personality.
- He proposed to Lady Bird on their first date, and after a whirlwind courtship, she accepted, beginning a partnership that lasted nearly four decades.

- Johnson once gave a late-night White House tour to a journalist, conducting it in his pajamas, revealing his informal side.

- He kept a three-way shower in the White House, designed to his specifications, reflecting his obsession with efficiency.

Ages at Death, Causes of Death, and Burial Locations

- Lyndon B. Johnson: Died at age 64 on January 22, 1973, of a heart attack at his Texas ranch. Buried at the LBJ Ranch Family Cemetery in Stonewall, Texas.

- Lady Bird Johnson: Died at age 94 on July 11, 2007, of natural causes in Austin, Texas. Buried alongside her husband at the LBJ Ranch Family Cemetery.

- Johnson's heavy smoking and stress from the presidency likely contributed to his heart issues, while Lady Bird's longevity reflected her active, healthy lifestyle.

Conclusion

Lyndon B. Johnson's life was a paradox of triumph and tragedy. His Great Society programs and civil rights achievements reshaped America, lifting millions out of poverty and advancing equality. Yet, his escalation of the Vietnam War fractured the nation and diminished his popularity. A man of immense energy and empathy, Johnson's legacy is a testament to the power of government to effect change, as well as the limits of leadership in times of crisis. Lady Bird's environmental and cultural contributions complemented his vision, leaving a lasting mark on the nation. Johnson's story reminds us that even flawed leaders can achieve greatness, but their mistakes carry heavy consequences.

Richard Nixon: The Watergate Scandal President

Description

Richard Nixon, the 37th President of the United States, was a complex figure whose legacy is intertwined with significant political achievements and the infamous Watergate scandal. His presidency was marked by foreign policy successes, domestic reforms, and a polarizing resignation that shaped his historical image.

Introduction

Richard Milhous Nixon served as president from January 20, 1969, to August 9, 1974, resigning amid the Watergate scandal, the only U.S. president to do so. His tenure included groundbreaking diplomatic efforts, particularly with China and the Soviet Union, alongside domestic policies addressing environmental and social issues. However, his legacy remains overshadowed by Watergate, a scandal involving illegal activities that eroded public trust in government.

Early Life

Nixon's early life was defined by humble beginnings and personal loss.

- He was born on January 9, 1913, in Yorba Linda, California.
- He grew up in a modest Quaker family that valued discipline and integrity.
- His childhood was marked by financial struggles and hardship.
- He experienced the loss of two brothers to tuberculosis.
- He excelled academically, showing a natural talent for debate and leadership.

Family

Nixon's family life and marriage reflected his grounding in Quaker values and resilience in the face of challenges.

- He was raised in a home rooted in Quaker traditions, which emphasized simplicity, discipline, and moral responsibility.
- His parents, Francis and Hannah Nixon, instilled a strong work ethic and sense of duty.
- He married Thelma Catherine "Pat" Ryan in 1940.
- Their marriage became a partnership that endured the demands of political life and personal challenges.

Children

Nixon and Pat had two daughters:

- Patricia "Tricia" Nixon, born February 21, 1946, who married Edward Cox in a White House ceremony in 1971.

- Julie Nixon, born July 5, 1948, who married Dwight David Eisenhower II, grandson of President Eisenhower, in 1968.

Rise to Power

Nixon's political career was marked by a rapid rise through the ranks and a reputation for strong anti-communist positions.

- In 1946, he won a seat in the U.S. House of Representatives.

- His firm stance against communism during the Alger Hiss case elevated his national profile.

- He was elected to the U.S. Senate in 1950.

- In 1953, he became Dwight D. Eisenhower's vice president, serving two terms.

- After losing the 1960 presidential race to John F. Kennedy, he made a political comeback.

- He won the presidency in 1968, returning to national leadership.

Influences

Nixon's worldview and political approach were shaped by personal values, historical study, and the Cold War era.

- His Quaker upbringing instilled ideals of peace, discipline, and integrity, though his political decisions sometimes diverged from these principles.
- Mentorship and collaboration with Dwight D. Eisenhower influenced his foreign policy strategies.
- The Cold War climate fueled his strong anti-communist convictions.
- His education in history and law sharpened his analytical and strategic thinking.

Party Affiliation

Nixon's political alignment reflected both ideological loyalty and pragmatic flexibility.

- He was a lifelong member of the Republican Party.
- He aligned with the party's conservative and anti-communist platforms.
- While primarily conservative, he embraced certain progressive domestic reforms when politically advantageous.
- His policies included measures like environmental protection, reflecting a willingness to adapt beyond strict partisan lines.

Presidency

Nixon's time in office was defined by both landmark achievements and historic scandals.

- He served as president from 1969 to 1974.

- His "Vietnamization" policy aimed to reduce U.S. troop involvement in Vietnam while strengthening South Vietnamese forces.

- He opened diplomatic relations with China, reshaping global geopolitics.

- He signed arms control agreements with the Soviet Union, promoting nuclear détente.

- Domestically, he established the Environmental Protection Agency and advanced school desegregation.

- His presidency ended in resignation following the Watergate scandal, which involved illegal political activities and a cover-up.

Accomplishments

- Ended U.S. involvement in the Vietnam War through Vietnamization.

- Opened diplomatic relations with China in 1972, a historic foreign policy shift.

- Signed the Strategic Arms Limitation Treaty (SALT I) with the Soviet Union.

- Established the Environmental Protection Agency (EPA) in 1970.

- Enforced desegregation in Southern schools, advancing civil rights.
- Created the Occupational Safety and Health Administration (OSHA).

First Lady's Contributions

- Promoted volunteerism, encouraging Americans to engage in community service.
- Supported historic preservation, notably restoring parts of the White House.
- Advocated for the arts, hosting cultural events to showcase American talent.
- Traveled internationally, enhancing U.S. soft power as a goodwill ambassador.

Positive Traits

Nixon's strengths were rooted in intelligence, persistence, and political skill.

- He demonstrated a deep understanding of global politics and strategic thinking.
- His resilience was evident in his political comeback after losing the 1960 presidential election and the 1962 California gubernatorial race.
- He maintained a strong work ethic throughout his career.
- He connected effectively with voters on economic issues, which bolstered his political success.

Negative Traits

Certain personal flaws overshadowed Nixon's accomplishments and damaged his legacy.

- His paranoia and distrust of opponents drove unethical behavior.
- He maintained a secretive nature and displayed vindictiveness toward critics.
- These tendencies alienated allies and eroded public trust.
- His priority on political survival over transparency contributed to the Watergate scandal.

Effects on Presidency

Nixon's strengths and weaknesses both left lasting marks on his time in office.

- His intelligence and strategic vision produced landmark foreign policy achievements, including détente with the Soviet Union and opening diplomatic relations with China.
- He applied his leadership to domestic reforms, such as establishing the Environmental Protection Agency.
- His paranoia and secrecy led directly to Watergate, undermining his accomplishments.
- The scandal forced his resignation and left a legacy of public distrust in government.

Pets

The Nixon family had several pets, most notably:

- Checkers, a cocker spaniel, gained fame during Nixon's 1952 "Checkers Speech," where he defended keeping the dog as a gift.
- Vicky, a poodle, and Pasha, a terrier, lived with the family in the White House.
- King Timahoe, an Irish setter, was a gift to Nixon during his presidency.

Religious Persuasion

Nixon's faith background influenced his values, but often contrasted with his political actions.

- He was raised as a Quaker, a faith emphasizing peace, simplicity, and integrity.
- While he maintained certain Quaker values, many of his political decisions—such as escalating the Vietnam War—diverged from these principles.
- He rarely spoke publicly about religion.
- In his youth, he regularly attended Quaker services.

Interesting Anecdotes

- In 1952, Nixon's "Checkers Speech" saved his vice-presidential candidacy by emotionally defending his family's modest finances and their dog, Checkers.

- During his 1959 Moscow visit, Nixon engaged in the impromptu "Kitchen Debate" with Soviet leader Nikita Khrushchev, showcasing American capitalism.

- Nixon, an avid musician, once played the piano on "The Tonight Show" in 1963, revealing a lesser-known personal side.

Ages at Death

- Richard Nixon died at age 81 on April 22, 1994.

- Pat Nixon died at age 81 on June 22, 1993.

Causes of Death

- Richard Nixon suffered a stroke and died from related complications.

- Pat Nixon died of lung cancer after years of declining health.

Burial Locations

- Richard Nixon was buried at the Richard Nixon Presidential Library and Museum in Yorba Linda, California.

- Pat Nixon was buried beside her husband at the Nixon Library in Yorba Linda, California.

Conclusion

Richard Nixon's life was a blend of remarkable achievements and profound failures. His foreign policy breakthroughs, including opening China and détente with the Soviet Union, reshaped global relations, while domestic reforms like the EPA left a lasting impact. Yet, the Watergate scandal defined his legacy, casting a shadow over his accomplishments and highlighting the consequences of his flaws. Pat Nixon's grace and contributions as First Lady complemented his tenure, though their personal lives were strained by political turmoil. Nixon's story remains a cautionary tale of ambition, power, and the fragility of public trust.

Gerald Ford: The Unelected President

Description

Gerald Rudolph Ford Jr., the 38th President of the United States, was a pragmatic leader who assumed the presidency during a time of national turmoil following Richard Nixon's resignation. Known for his integrity and steady demeanor, Ford's tenure was marked by efforts to heal a divided nation and stabilize the economy, though his controversial pardon of Nixon defined much of his legacy. His presidency, from 1974 to 1977, remains notable for its unique context as the only president never elected to either the presidency or vice presidency.

Introduction

Gerald Ford's ascent to the presidency was unprecedented, stepping into the role after serving as vice president under Richard Nixon, who resigned amid the Watergate scandal. A career politician with a reputation for decency, Ford prioritized restoring public trust in government. His brief presidency faced challenges like inflation, the aftermath of Vietnam, and Cold War tensions. While often overshadowed by the pardon of Nixon, Ford's leadership provided stability during a critical period in American history.

Early Life

Gerald Ford was born Leslie Lynch King Jr. on July 14, 1913, in Omaha, Nebraska. His early years shaped his grounded character and commitment to public service.

- Moved to Grand Rapids, Michigan, as a toddler after his parents' divorce.
- Adopted by his stepfather, Gerald Rudolff Ford, whose name he took.
- Excelled in athletics, particularly football, at Grand Rapids South High School.
- Attended the University of Michigan, where he played center and linebacker, contributing to two national championship teams (1932–1933).
- Earned a law degree from Yale University in 1941, working as a football coach to support his studies.

Family

Ford's family life was rooted in Midwestern values, emphasizing loyalty and community.

- Son of Dorothy Ayer Gardner and Leslie Lynch King Sr. (biological father).

- Raised by stepfather Gerald Rudolff Ford, a paint salesman, whom he considered his true father.
- Close relationship with his mother, who instilled resilience and discipline.
- Maintained strong ties with his three half-brothers from his mother's remarriage.

Children

Gerald and Betty Ford had four children, who grew up in a supportive but public family environment due to their father's political career.

- Michael Gerald Ford (born 1950) became a minister.
- John "Jack" Gardner Ford (born 1952) pursued a career in business and journalism.
- Steven Meigs Ford (born 1956) worked as an actor and rancher.
- Susan Elizabeth Ford (born 1957) became a photographer and advocate.

Rise to Power

Ford's political career was marked by steady progression through dedication and bipartisan respect. v

- Served in the U.S. Navy during World War II (1942–1946), earning multiple commendations aboard the USS Monterey.
- Elected to the U.S. House of Representatives in 1948, representing Michigan's 5th District for 25 years.

- Rose to House Minority Leader in 1965, known for his collegiality and legislative skill.
- Appointed vice president in 1973 under Nixon after Spiro Agnew's resignation.
- Became president on August 9, 1974, following Nixon's resignation.

Influences

Ford's leadership was shaped by his experiences and mentors.

- His stepfather's work ethic and integrity influenced his approach to public service.
- Exposure to bipartisan politics in Congress fostered his belief in compromise.
- The Watergate scandal and national distrust in government underscored his focus on transparency.
- Admired historical figures like Abraham Lincoln for their unifying leadership.

Party Affiliation

Ford was a lifelong Republican, aligning with the party's moderate wing.

- Advocated fiscal conservatism, supporting balanced budgets and limited government spending.
- Embraced social moderation, avoiding the ideological extremes of his era.

- Worked across party lines, earning respect from Democrats and Republicans alike.

Presidency

Ford's presidency (1974–1977) focused on restoring trust and addressing economic challenges.

- Assumed office amid the Watergate fallout and economic stagflation.

- Issued a controversial pardon of Richard Nixon in September 1974 to move the nation forward.

- Faced high inflation and unemployment, implemented the WIN (Whip Inflation Now) program.

- Oversaw the final withdrawal of U.S. forces from Vietnam in 1975.

- Signed the Helsinki Accords, easing Cold War tensions with the Soviet Union.

- Survived two assassination attempts in September 1975, demonstrating resilience.

Accomplishments

- Restored public confidence in the presidency through transparency and accessibility.

- Stabilized U.S. foreign policy by maintaining détente with the Soviet Union.

- Signed the Education for All Handicapped Children Act (1975), advancing special education.

- Promoted energy conservation to address the 1970s energy crisis.
- Facilitated the safe evacuation of Americans from Saigon during Vietnam's fall.

First Lady's Contributions

Betty Ford, as First Lady, was a trailblazer for her candor and advocacy.

- Publicly discussed her breast cancer diagnosis, raising awareness and reducing stigma.
- Advocated for the Equal Rights Amendment, supporting women's rights.
- Founded the Betty Ford Center for addiction treatment after her own struggles with alcoholism.
- Promoted arts and culture, hosting events to showcase American talent.
- Championed mental health awareness, influencing national conversations.

Positive Traits

Ford's personal qualities shaped his presidency positively.

- Integrity: Known as "Mr. Nice Guy," his honesty helped rebuild trust post-Watergate.
- Pragmatism: Approached problems with practical solutions, avoiding ideological rigidity.
- Humility: Remained approachable, often engaging directly with the public and press.

- These traits fostered a calming presence, though some criticized his lack of bold vision.

Negative Traits

Ford's flaws occasionally hindered his leadership.

- Indecisiveness: Critics noted his hesitancy on complex issues like the economy.
- Loyalty: His loyalty to Nixon, evident in the pardon, alienated voters.
- Lack of charisma: His reserved demeanor struggled to inspire a weary public.
- These traits contributed to his 1976 election loss to Jimmy Carter.

Pets

The Fords brought warmth to the White House with their pets.

- Liberty, a golden retriever, was a beloved companion, often photographed with the family.
- Liberty gave birth to puppies in the White House, delighting staff and visitors.
- The Fords also had a Siamese cat named Shan, less publicized but part of the household.

Religious Persuasion

Ford was a devout Episcopalian whose faith guided his moral compass.

- Regularly attended services at St. John's Episcopal Church near the White House.
- His faith emphasized forgiveness, influencing his decision to pardon Nixon.
- Kept his religious beliefs private, avoiding public displays of piety.

Interesting Anecdotes

Ford's life was filled with memorable moments that highlighted his character.

- As a young man, he turned down NFL offers from the Green Bay Packers and Detroit Lions to pursue law.
- Tripped on Air Force One steps in 1975, leading to a media caricature as clumsy, despite his athletic background.
- During a 1975 debate preparation, he humorously practiced dodging questions about the Nixon pardon.
- Betty Ford's candid TV interview about her mastectomy shocked but inspired millions.

Ages at Death

- Gerald Ford died at age 93 on December 26, 2006.
- Betty Ford died at age 93 on July 8, 2011.

Causes of Death

- Gerald Ford succumbed to heart disease and pneumonia at his home in Rancho Mirage, California.

- Betty Ford passed away from natural causes at Eisenhower Medical Center in Rancho Mirage.

Burial Locations

- Gerald Ford is buried at the Gerald R. Ford Presidential Museum in Grand Rapids, Michigan.

- Betty Ford is buried alongside her husband at the Gerald R. Ford Presidential Museum.

Conclusion

Gerald Ford's presidency, though brief and unelected, was a pivotal chapter in American history. His steady leadership helped navigate the nation through the Watergate aftermath, economic challenges, and the end of the Vietnam War. While the Nixon pardon remains his most controversial act, it reflected his belief in national healing over division. Betty Ford's courageous advocacy complemented his tenure, leaving a lasting impact on health and social issues. Ford's legacy as a decent, pragmatic leader endures, remembered for restoring stability when America needed it most.

Jimmy Carter: Champion of Peace and Humanitarianism

Introduction

James Earl Carter Jr., the 39th President of the United States, was a man defined by his commitment to peace, human rights, and humanitarian work. Born in a small rural town, Carter rose from humble beginnings to become a global figure, leaving a legacy that extends far beyond his single term in office (1977–1981). His presidency faced significant challenges, including economic turmoil and international crises, but his post-presidency redefined the role of a former president through his tireless efforts in diplomacy, philanthropy, and advocacy. Awarded the Nobel Peace Prize in 2002, Carter's life reflects a blend of moral conviction, resilience, and dedication to service, making him one of the most impactful former presidents in U.S. history.

Early Life

Jimmy Carter was born on October 1, 1924, in Plains, Georgia, a small farming community. His early years were shaped by rural life, hard work, and a strong sense of community.

- Grew up in Archery, a rural area near Plains, in a home without electricity or indoor plumbing until he was 11.
- Son of James Earl Carter Sr., a businessman and peanut farmer, and Lillian Gordy Carter, a nurse who defied racial norms by treating Black patients.
- Worked on the family peanut farm from a young age, selling produce in town and saving money to buy rental properties by age 13.
- Attended all-white Plains High School, where he was a diligent student with a love for reading and basketball.
- Developed an early interest in woodworking through Future Farmers of America.
- Enrolled at Georgia Southwestern College (1941–1942), transferred to Georgia Tech (1942–1943), and graduated from the U.S. Naval Academy in 1946.

Family

Carter's family life was rooted in Plains, Georgia, and centered around his lifelong partnership with Rosalynn Smith, whom he married in 1946.

- Married Rosalynn Smith on July 7, 1946, a childhood acquaintance from Plains.

- Parents, Earl and Lillian Carter, were influential figures; Earl was a community leader, and Lillian was a progressive nurse and later a Peace Corps volunteer.
- Had three siblings: sisters Gloria and Ruth, and brother Billy, who later gained attention for controversial business dealings.
- The Carters maintained close ties to Plains, returning there after his presidency to live a modest life.

Children

Jimmy and Rosalynn Carter had four children, who grew up in the public eye during Jimmy's political career.

- John William ("Jack") Carter, born in 1947, pursued business and politics, running for U.S. Senate in Nevada in 2006.
- James Earl III ("Chip") Carter, born in 1950, worked in the family business and was involved in Democratic politics.
- Donnel Jeffrey ("Jeff") Carter, born in 1952, kept a lower profile, focusing on family and business.
- Amy Lynn Carter, born in 1967, was a child during the presidency, drawing media attention; later, she became an artist and illustrated one of her father's books.

Rise to Power

Carter's ascent to the presidency was remarkable for a relatively unknown Southern governor, driven by his determination and strategic campaigning.

- Served in the Georgia State Senate (1963–1967), focusing on education and exposing election fraud to secure his seat.

- Ran for governor of Georgia in 1966 but lost to segregationist Lester Maddox; won in 1970, serving as governor from 1971 to 1975.

- As governor, he declared "the time for racial discrimination is over," increasing Black representation in state government by 25 percent and reorganizing agencies for efficiency.

- Announced his presidential candidacy in 1974, positioning himself as a Washington outsider post-Watergate.

- Published: *Why Not the Best?* (1975) to introduce himself to voters, emphasizing honesty and integrity.

- Won the Democratic nomination in 1976 as a dark horse and narrowly defeated incumbent Gerald Ford, with 297 electoral votes to Ford's 241.

Influences

Carter's values and policies were shaped by a blend of personal experiences, religious faith, and key figures.

- Influenced by his mother Lillian's progressive views on race and service, which shaped his civil rights stance.

- Inspired by a sermon asking, "If you were arrested for being a Christian, would there be enough evidence to convict you?" prompted lifelong reflection on faith and action.

- Admiral Hyman Rickover, Carter's Navy mentor, instilled discipline and high standards during his work on nuclear submarines.
- The Watergate scandal and Vietnam War fueled his campaign as an honest outsider, resonating with a disillusioned electorate.
- Woodrow Wilson's progressive Southern tradition influenced his focus on reform and human rights.

Party Affiliation

Carter was a lifelong member of the Democratic Party, though his approach often diverged from traditional party lines.

- Ran as a moderate Democrat in the Georgia State Senate, countering the state's segregationist faction.
- As governor and president, he promoted progressive policies like civil rights and government efficiency, but clashed with liberal Democrats over fiscal restraint.
- His emphasis on human rights and deregulation aligned with emerging centrist Democratic ideals, paving the way for the party's evolution in the 1990s.

Presidency

Carter's presidency (January 20, 1977–January 20, 1981) was marked by ambitious reforms, international diplomacy, and significant challenges like stagflation and the Iran hostage crisis.

- Took office during economic "stagflation" (high inflation and slow growth), focusing on reducing deficits and government spending.

- Faced the 1979 energy crisis, promoting conservation and alternative energy, though his policies were complex and poorly understood by the public.

- Struggled with Congress due to his outsider status and reluctance to engage in political deal-making, leading to tensions with Democratic leaders like Speaker Tip O'Neill.

- Handled the Iran hostage crisis (1979–1981), which dominated his final year and contributed to his 1980 election loss to Ronald Reagan.

- Continued negotiations for hostage release, with the 52 Americans freed minutes after his term ended.

Accomplishments

- **Camp David Accords (1978):** Mediated a historic peace treaty between Egypt and Israel, ending decades of conflict and stabilizing the Middle East.

- **Panama Canal Treaties (1977):** Secured ratification to return the canal to Panama, honoring U.S. commitments and improving Latin American relations.

- **SALT II Treaty (1979):** Signed with the Soviet Union to limit nuclear arms, though ratification was stalled by the Soviet invasion of Afghanistan.

- **Department of Energy and Education (1977, 1979):** Established these cabinet-level departments to address energy policy and expand education funding.

- **Energy Policy:** Reduced U.S. oil imports from 48% to 40% by 1980, doubled coal production, and promoted sustainable energy like wind and solar.

- **Deregulation:** Deregulated airlines, trucking, and railroads, lowering costs and fostering economic flexibility.

- **Environmental Protection:** Created the Superfund to clean up toxic waste sites and protected 100 million acres of Alaskan land.

- **Human Rights:** Made human rights a cornerstone of U.S. foreign policy, challenging allies and adversaries alike.

- **Civil Service Reform:** Reorganized federal agencies for efficiency and accountability.

- **Vietnam Draft Evaders Amnesty (1977):** Issued Proclamation 4483, granting unconditional amnesty to Vietnam War draft evaders.

First Lady's Contributions

Rosalynn Carter redefined the role of First Lady, serving as an active partner in policy and diplomacy.

- **Mental Health Advocacy:** Chaired hearings on mental health in 1978, pushing for reform and reducing stigma around mental illness.

- **Diplomatic Missions:** Represented the U.S. in Latin America in 1977, meeting leaders in seven countries.

- **Active Advisor:** Attended cabinet meetings and advised the president on key issues, earning recognition as a full partner in governance.

- **Carter Center Co-Founder:** Co-established The Carter Center in 1982, contributing to global health, democracy, and human rights initiatives.

- **Habitat for Humanity:** Worked alongside Jimmy Carter on housing projects, becoming a prominent figure in the organization's mission.

- **Author and Advocate:** Wrote *First Lady from Plains* (1984) and continued advocacy for caregiving and mental health post-presidency.

Personal Traits

Carter's personality shaped both his successes and challenges as president.

Positive Traits

- **Integrity:** Known for honesty and moral conviction, earning trust post-Watergate.

- **Work Ethic:** Diligent and detail-oriented, mastering complex policy issues.

- **Compassion:** Deeply empathetic, driven by a desire to serve the public.

- **Idealism:** Committed to human rights and peace, prioritizing global good over political expediency.

Negative Traits

- **Stubbornness:** Refused to compromise on key issues, alienating Congress and hindering legislative success.

- **Micromanagement:** Over-involvement in details led to perceptions of disorganization.

- **Aloofness:** Struggled with backroom politicking, straining relations with allies.

- **Pessimistic Tone:** His "malaise" speech (1979), though not using the term, was perceived as gloomy, eroding public confidence.

Effects on Presidency

Carter's integrity and idealism drove landmark achievements like the Camp David Accords, but his stubbornness and aloofness led to a rocky relationship with Congress, limiting domestic policy success. His focus on complex issues like energy policy was forward-thinking but poorly communicated, contributing to his 1980 election loss.

Pets

The Carters had a few pets during their White House years, reflecting their down-to-earth nature.

- **Grits:** A border collie mix, gifted to daughter Amy by her teacher, named after a Southern dish.

- **Misty Malarky Ying Yang:** A Siamese cat belonging to Amy, known for roaming the White House grounds.

- **Lewis Brown:** A hound dog, named after a family friend, kept at their Plains home.

Religious Persuasion

Carter's deep faith profoundly influenced his life and leadership.

- Evangelical Baptist, raised in Plains Baptist Church, where he served as a deacon and Sunday school teacher.
- Popularized the term "born again" during his 1976 campaign, emphasizing personal faith.
- Prayed daily as president, citing Jesus as the driving force in his life.
- Renounced Southern Baptist Convention membership in 2000 over its stance on women, joining the Cooperative Baptist Fellowship.
- Taught Sunday school at Maranatha Baptist Church in Plains until 2019, earning a posthumous Grammy in 2025 for *Last Sundays in Plains*.

Interesting Anecdotes

Carter's life is filled with unique stories that highlight his character.

- As a young naval officer, Carter refused to attend a segregated party in the Bahamas, leading his entire submarine crew to boycott the event.

- During a 1979 fishing trip, Carter was famously "attacked" by a swamp rabbit, a story exaggerated by the media but reflective of his unpretentious nature.
- At 13, Carter bought five houses during the Great Depression, showcasing an early entrepreneurial spirit.
- He and Rosalynn walked from the Capitol to the White House during the 1977 inaugural parade, breaking tradition to connect with the public.
- Carter cast a mail ballot for Kamala Harris in October 2024, marking his 21st presidential election as a voter.

Ages at Death

- **Jimmy Carter:** Died at age 100 on December 29, 2024, the longest-lived U.S. president.
- **Rosalynn Carter:** Died at age 96 on November 19, 2023.

Causes of Death

- **Jimmy Carter:** Died of natural causes after over a year in hospice care at his home in Plains, Georgia.
- **Rosalynn Carter:** Died of natural causes, compounded by dementia, diagnosed in May 2023, at their Plains home.

Burial Locations

- **Jimmy Carter:** Buried in Plains, Georgia, with funeral ceremonies held in Plains and Washington, D.C., reflecting his roots and national service.
- **Rosalynn Carter:** Buried in Plains, Georgia, alongside her husband, in a private family plot.

Conclusion

Jimmy Carter's life was a testament to the power of integrity, faith, and service. His presidency, though challenged by economic and international crises, achieved lasting successes like the Camp David Accords and energy reforms that shaped modern policy. His partnership with Rosalynn Carter redefined the roles of president and first lady, emphasizing collaboration and advocacy. Post-presidency, Carter's work through The Carter Center and Habitat for Humanity cemented his legacy as a global peacemaker and humanitarian, earning him the 2002 Nobel Peace Prize. Despite political setbacks, his commitment to human rights, environmental protection, and compassion left an indelible mark. Carter's death at 100, following Rosalynn's at 96, marked the end of a remarkable era, but their legacy of service endures in Plains and beyond.

Ronald Reagan: The Great Communicator

Description

Ronald Reagan, the 40th President of the United States, served from 1981 to 1989. Known for his charisma, optimism, and ability to connect with the American public, Reagan earned the nickname "The Great Communicator." His presidency focused on economic reform, a strong anti-communist stance, and restoring national pride.

Introduction

Ronald Reagan's life spanned a remarkable journey from a small-town upbringing to Hollywood stardom, and ultimately to the White House. His leadership style, rooted in clear communication and conservative principles, reshaped American politics and global relations during the Cold War era. This summary explores his life, legacy, and the contributions of his wife, Nancy Reagan.

Early Life

Reagan was born in humble circumstances and developed a strong work ethic and optimistic outlook that shaped his later years. His early experiences in radio and acting laid the foundation for his public persona.

- Born on February 6, 1911, in Tampico, Illinois, to John "Jack" Reagan, a shoe salesman, and Nelle Wilson Reagan, a homemaker and devout Christian.
- Grew up in Dixon, Illinois, where he worked as a lifeguard, reportedly saving 77 lives.
- Attended Eureka College, earning a degree in economics and sociology in 1932.
- Began his career as a radio sports announcer, showcasing his distinctive voice and communication skills.

Family

Reagan's family life was marked by two marriages and a close partnership with his second wife, Nancy, who played a significant role in his personal and political life.

- First marriage to actress Jane Wyman (1940–1948); divorced due to differing career paths and personal strains.
- Second marriage to Nancy Davis (1952–2004), a former actress who became his closest confidante.
- Maintained close relationships with his children, though family dynamics were sometimes strained due to his demanding career.

Children

Reagan had four children, two from each marriage, who pursued varied paths in life.

- Maureen Reagan (1941–2001), from his marriage to Jane Wyman, became a political activist and author.

- Michael Reagan (born 1945), adopted during Reagan's marriage to Wyman, worked as a radio host and conservative commentator.

- Patricia "Patti" Davis (born 1952), from his marriage to Nancy, pursued acting and writing, often critical of her father's politics.

- Ronald "Ron" Prescott Reagan (born 1958), from his marriage to Nancy, became a journalist and liberal commentator.

Rise to Power

Reagan's transition from entertainment to politics was gradual but deliberate, driven by his growing conservative convictions and public speaking prowess.

- Gained fame as a Hollywood actor in the 1930s and 1940s, starring in films like *Knute Rockne, All-American,* and *Kings Row*.

- Served as president of the Screen Actors Guild (1947–1952, 1959–1960), where he honed leadership skills and confronted communism in Hollywood.

- Became a corporate spokesman for General Electric in the 1950s, delivering speeches that refined his conservative ideology.

- Elected governor of California (1967–1975), where he implemented welfare reforms and gained national prominence.
- Ran for president in 1976, narrowly losing the Republican nomination to Gerald Ford, before winning in 1980 against Jimmy Carter.

Influences

Reagan's worldview was shaped by personal experiences, historical events, and key figures who reinforced his belief in individual liberty and limited government.

- His mother, Nelle, instilled strong Christian values and a sense of optimism.
- Inspired by conservative thinkers like Barry Goldwater and economist Milton Friedman, who advocated free-market principles.
- The Cold War and Soviet expansionism fueled his staunch anti-communist stance.
- His time in Hollywood exposed him to the dangers of collectivism, shaping his distrust of centralized power.

Party Affiliation

Reagan's political evolution reflected a shift from liberal to conservative ideals, aligning with the Republican Party's growing conservative wing.

- Initially a Democrat, supporting Franklin D. Roosevelt and the New Deal in the 1930s and 1940s.

- Shifted to the Republican Party in the early 1960s, driven by concerns over government overreach and communism.
- Became a leading figure in the conservative movement, advocating for lower taxes, deregulation, and a strong national defense.

Presidency

Reagan's presidency (1981–1989) was defined by economic reforms, foreign policy triumphs, and domestic challenges. His leadership style emphasized optimism and decisiveness, though controversies arose.

- Elected in 1980 in a landslide against Jimmy Carter, capitalizing on economic discontent and a call for national renewal.
- Survived an assassination attempt in 1981, which bolstered his public image as resilient and relatable.
- Pursued "Reaganomics," a policy of tax cuts, deregulation, and reduced government spending to stimulate the economy.
- Took a hardline stance against the Soviet Union, calling it the "Evil Empire" and escalating the arms race, contributing to its eventual collapse.
- Faced challenges, including the Iran-Contra scandal, which tarnished his administration's reputation but did not lead to his impeachment.

Accomplishments

Reagan's presidency left a lasting impact on the economy, foreign policy, and American culture.

- Implemented the Economic Recovery Tax Act of 1981, reducing tax rates and spurring economic growth.

- Appointed three Supreme Court justices, including Sandra Day O'Connor, the first woman to serve on the Court.

- Strengthened U.S. military through defense spending, projecting power during the Cold War.

- Negotiated with Soviet leader Mikhail Gorbachev, leading to the INF Treaty (1987), reducing nuclear arsenals.

- Restored national confidence with his optimistic rhetoric and vision of America as a "shining city on a hill."

First Lady's Contributions

Nancy Reagan played a significant role as First Lady, focusing on social issues and supporting her husband's agenda.

- Launched the "Just Say No" campaign to combat drug abuse, raising awareness among youth and families.

- Influenced White House staffing and policy decisions, acting as a trusted advisor to the president.

- Oversaw White House renovations, restoring elegance and historical integrity to the residence.

- Advocated for Alzheimer's research after Reagan's diagnosis, founding the Ronald and Nancy Reagan Research Institute.

Positive Traits and Their Effects

Reagan's personal qualities shaped his presidency, enhancing his ability to lead and connect with the public.

- Charisma and communication skills earned him the "Great Communicator" moniker, making complex policies accessible and inspiring public support.
- Optimism and resilience, evident after the assassination attempt, bolstered national morale and his leadership image.
- Conviction in conservative principles guided his policy agenda, appealing to a broad coalition of voters.

Negative Traits and Their Effects

Reagan's weaknesses led to challenges during his presidency, though his popularity often mitigated their impact.

- Detached management style sometimes resulted in oversight failures, notably during the Iran-Contra scandal.
- Reluctance to address social issues like AIDS early in his presidency drew criticism for insensitivity.
- Overemphasis on deregulation contributed to economic disparities, raising concerns about income inequality.

Pets

Reagan's love for animals, particularly dogs, reflected his warmth and relatability, endearing him to the public.

- Owned a Cavalier King Charles Spaniel named Rex, a gift from Nancy, who often accompanied the Reagans at the White House.
- Kept horses at their California ranch, including a favorite named El Alamein, reflecting Reagan's passion for riding.
- Other pets included cats and dogs at their ranch, emphasizing their rural, down-to-earth lifestyle.

Religious Persuasion

Reagan's faith informed his moral outlook and public rhetoric, though he was not overtly religious in his personal practice.

- Raised in the Disciples of Christ church by his devout mother, Nelle.
- Maintained a general Christian faith, often referencing God and biblical themes in speeches.
- Rarely attended church during his presidency due to security concerns, but expressed a belief in divine providence guiding America.

Interesting Anecdotes

Reagan's life was filled with colorful stories that highlighted his humor, charm, and resilience.

- As a lifeguard, he notched a stick for each life saved, totaling 77 by the end of his tenure.
- During his Hollywood days, he once ad-libbed a radio broadcast of a baseball game, inventing details when the feed cut out.
- After surviving the 1981 assassination attempt, he quipped to Nancy, "Honey, I forgot to duck," showcasing his humor under pressure.
- He famously challenged Gorbachev to "tear down this wall" in a 1987 Berlin speech, a moment credited with galvanizing the end of the Cold War.

Ages at Death, Causes of Death, and Burial Locations

Reagan and Nancy lived long lives but faced health challenges in their later years.

- Ronald Reagan died on June 5, 2004, at age 93, from pneumonia complicated by Alzheimer's disease; buried at the Ronald Reagan Presidential Library in Simi Valley, California.
- Nancy Reagan died on March 6, 2016, at age 94, from congestive heart failure; buried beside Ronald at the Reagan Presidential Library.

Conclusion

Ronald Reagan's legacy as the Great Communicator endures through his transformative presidency, which reshaped American conservatism and contributed to the end of the Cold War. His optimism, clear vision, and partnership with Nancy Reagan defined an era of renewed national pride. Despite controversies, his ability to inspire and lead left an indelible mark on history, making him one of America's most influential presidents.

George H. W. Bush: Architect of the New World Order

Description

George Herbert Walker Bush, the 41st President of the United States, was a statesman, diplomat, and public servant whose career spanned decades of service in various roles, including Vice President, CIA Director, and Ambassador to the United Nations. Known for his pragmatic leadership and foreign policy expertise, Bush navigated the end of the Cold War and led the U.S. through significant global and domestic challenges.

Introduction

George H. W. Bush's life was marked by a commitment to public service, shaped by his experiences in World War II, his political career, and his dedication to family. His presidency (1989–1993) is most remembered for his role in shaping a "New World Order," particularly through his leadership in the Gulf War and the collapse of the Soviet

Union. His steady, consensus-driven approach defined his legacy as a leader who prioritized stability and diplomacy.

Early Life

Born on June 12, 1924, in Milton, Massachusetts, George H. W. Bush grew up in a privileged and politically connected family. His early years were marked by a strong sense of duty and ambition.

- Raised in Greenwich, Connecticut, in a wealthy, well-educated household.
- Attended Phillips Academy in Andover, Massachusetts, where he excelled academically and athletically.
- Enlisted in the U.S. Navy at 18, becoming one of the youngest naval aviators in World War II.
- Flew 58 combat missions, earning the Distinguished Flying Cross for bravery after being shot down in 1944.

Family

Bush's family life was central to his identity, with a strong partnership with his wife, Barbara, and a large, close-knit family.

- Married Barbara Pierce on January 6, 1945, after meeting her at a dance during his Navy service.
- The couple moved frequently early in their marriage, settling in Texas to build a life in the oil industry.
- Their marriage lasted 73 years, one of the longest of any presidential couple, marked by mutual support and shared values.

Children

George and Barbara Bush had six children, five of whom survived to adulthood, with two becoming prominent political figures.

- George W. Bush (born 1946), 43rd President of the United States.
- Robin Bush (1949–1953) died of leukemia at age three, deeply affecting the family.
- John "Jeb" Bush (born 1953), former Governor of Florida and presidential candidate.
- Neil Bush (born 1955), businessman.
- Marvin Bush (born 1956), businessman.
- Dorothy "Doro" Bush Koch (born 1959), philanthropist.

Rise to Power

Bush's ascent to the presidency was a steady climb through various political and diplomatic roles, showcasing his versatility and ambition.

- Moved to Texas in the 1940s, founding a successful oil company, Zapata Petroleum Corporation.
- Elected to the U.S. House of Representatives from Texas in 1966, serving two terms.
- Appointed Ambassador to the United Nations (1971–1973), enhancing his foreign policy credentials.
- Served as Chairman of the Republican National Committee (1973–1974) during the Watergate scandal.

- Director of the CIA (1976–1977), restoring morale after congressional investigations.
- Vice President under Ronald Reagan (1981–1989), playing a key role in foreign policy.
- Won the presidency in 1988, defeating Michael Dukakis in a landslide.

Influences

Bush's worldview was shaped by his upbringing, military service, and political mentors.

- His father, Prescott Bush, a U.S. Senator, instilled a sense of public service and moderation.
- World War II experience fostered a commitment to duty and internationalism.
- Ronald Reagan's conservative policies influenced Bush's domestic agenda, though he leaned more pragmatic.
- His New England roots and Texas connections blended patrician values with populist appeal.

Party Affiliation

- Republican, aligning with the moderate wing of the party.
- Advocated fiscal conservatism, international engagement, and pragmatic governance.
- Initially supported Reagan's supply-side economics but later distanced himself from his "no new taxes" pledge, which he famously broke.

Presidency

Bush's presidency (1989–1993) was defined by foreign policy triumphs and domestic challenges, navigating a rapidly changing global landscape.

- Led the U.S. in the Gulf War (1991), assembling a coalition to expel Iraqi forces from Kuwait.
- Oversaw the end of the Cold War, managing U.S.-Soviet relations as the USSR collapsed.
- Signed the Americans with Disabilities Act (1990), a landmark civil rights law.
- Faced economic recession and criticism for breaking his tax pledge, contributing to his 1992 election loss to Bill Clinton.
- Promoted a "kinder, gentler" conservatism, emphasizing volunteerism and community service.

Accomplishments

- Successfully led the Gulf War, restoring Kuwait's sovereignty with minimal U.S. casualties.
- Negotiated the Strategic Arms Reduction Treaty (START I) with the Soviet Union, reducing nuclear arsenals.
- Signed the Clean Air Act Amendments (1990), addressing acid rain and urban air pollution.
- Championed the North American Free Trade Agreement (NAFTA), laying the groundwork for its passage under Clinton.

- Established the Points of Light Foundation, promoting volunteerism.

First Lady's Contributions

Barbara Bush was a beloved First Lady, known for her warmth and advocacy for literacy.

- Founded the Barbara Bush Foundation for Family Literacy, promoting reading programs nationwide.

- Advocated for education and children's welfare, visiting schools and libraries.

- Supported AIDS awareness, visiting patients to reduce stigma.

- Authored books, including *Millie's Book*, with proceeds benefiting literacy causes.

- Her candid, approachable style made her a popular figure, enhancing Bush's public image.

Positive Traits

- Pragmatic and diplomatic, excelling in coalition-building and crisis management.

- Deeply loyal, maintaining lifelong friendships and a strong family bond.

- Disciplined and hardworking, with a focus on duty and service.

- Resilient, recovering from personal tragedies like the loss of his daughter Robin.

Negative Traits

- Perceived as aloof or out of touch, struggling to connect with voters on domestic issues.
- Indecisiveness on economic policy, particularly his tax pledge reversal, damaged credibility.
- Reserved demeanor sometimes hindered his ability to inspire or rally the public.

Effects of Traits on Presidency

- Positive: His diplomatic skills fostered international alliances, critical for the Gulf War and Cold War resolution. His loyalty built trust among advisors, ensuring effective governance.
- Negative: His aloofness and tax pledge reversal alienated conservative voters, contributing to his 1992 defeat. His reserved nature limited his ability to counter Clinton's charismatic campaign.

Pets

- Millie, a springer spaniel, was a beloved White House pet, featured in Barbara's book.
- Ranger, another springer spaniel, was a gift from Millie's litter, often accompanying Bush on walks.

Religious Persuasion

- Episcopalian, raised in a devout family with strong Christian values.
- Attended church regularly but kept faith private, avoiding public displays of religiosity.

- His faith influenced his emphasis on family, duty, and compassion in public life.

Interesting Anecdotes

- Survived being shot down in the Pacific during WWII, rescued by a submarine after hours at sea.
- Famously vomited on Japanese Prime Minister Kiichi Miyazawa during a 1992 state dinner due to illness.
- Skydived on his 80th, 85th, and 90th birthdays, fulfilling a lifelong passion.
- Maintained a tradition of writing personal thank-you notes, reflecting his courteous nature.

Ages at Death

- George H. W. Bush died on November 30, 2018, at age 94.
- Barbara Bush died on April 17, 2018, at age 92.

Causes of Death

- George H. W. Bush: Died of complications from Parkinson's disease and old age.
- Barbara Bush: Died of chronic obstructive pulmonary disease and congestive heart failure.

Burial Locations

- Both George and Barbara Bush are buried at the George Bush Presidential Library and Museum in College Station, Texas, alongside their daughter Robin.

Conclusion

George H. W. Bush's life was a testament to duty, resilience, and diplomacy. His presidency, though marred by domestic economic struggles, left a lasting mark on global affairs through his leadership in the Gulf War and the end of the Cold War. Barbara Bush complemented his legacy with her advocacy for literacy and her relatable persona. Together, they embodied a commitment to service, leaving a legacy of pragmatism and compassion that continues to influence American politics.

Bill Clinton: The Comeback Kid

Description

Bill Clinton, the 42nd President of the United States, served from 1993 to 2001. Known for his charisma and ability to connect with people, he navigated a tumultuous political landscape marked by economic prosperity and personal scandals.

Introduction

William Jefferson Clinton, often called the "Comeback Kid" for his resilience in overcoming political setbacks, led the U.S. during a period of significant economic growth and global change. His presidency is remembered for its focus on centrist policies, international diplomacy, and controversies that shaped his legacy.

Early Life

Born on August 19, 1946, in Hope, Arkansas, William Jefferson Blythe III grew up in modest circumstances. His early life was shaped by family changes and a growing interest in learning and leadership.

- His father died before his birth, leaving his mother to raise him.
- His mother's remarriage played an important role in shaping his early years.
- He excelled academically throughout his youth.
- He also showed an early interest in politics.

Family

Clinton's family life centered around his mother, Virginia Kelley, and his stepfather, Roger Clinton, whose name he later adopted. His upbringing in Hot Springs, Arkansas, reflected both warmth and challenges within the household.

- He was raised primarily by his mother, Virginia Kelley.
- He took the surname of his stepfather, Roger Clinton.
- His childhood in Hot Springs, Arkansas, was shaped by a complex family dynamic.
- His stepfather's struggles with alcoholism created difficulties in the household.

Children

Bill and Hillary Clinton have one daughter, Chelsea Clinton, who was born on February 27, 1980. She grew up in the public eye and went on to build a career dedicated to service and leadership.

- Chelsea Clinton was born on February 27, 1980.
- She was raised in the spotlight as the child of prominent political figures.
- She later pursued a career in public service.
- She became actively involved in philanthropy and global initiatives.

Rise to Power

Clinton's political career began early, shaped by his strong academic background and rapid rise in public service. Before becoming president in 1992, he built a reputation as a pragmatic Democrat in Arkansas politics.

- He graduated from Georgetown University.
- He studied as a Rhodes Scholar at Oxford.
- He earned a law degree from Yale University.
- He served as Arkansas Attorney General from 1977 to 1979.
- He was Governor of Arkansas from 1979 to 1981 and again from 1983 to 1992.
- He won the presidency in 1992.

Influences

Clinton drew inspiration from prominent leaders and his own upbringing, which together shaped his political vision. His influences combined ideals of service, equality, and moderation.

- He was inspired by John F. Kennedy, whose charisma fueled his political ambitions.
- He was influenced by Martin Luther King Jr., whose civil rights advocacy shaped his views on equality.
- His Southern roots guided his understanding of regional and cultural perspectives.
- His moderate political stance influenced his approach to governance.

Party Affiliation

Clinton remained a lifelong Democrat, shaping his political identity within the party's evolving landscape. His philosophy reflected a centrist approach that balanced progressive goals with fiscal responsibility.

- He was a lifelong member of the Democratic Party.
- He aligned with the party's "New Democrat" wing.
- He promoted centrist policies that blended progressive ideals with pragmatism.
- He emphasized fiscal responsibility while advancing social progress.

Presidency

Clinton's presidency (1993–2001) was defined by a mix of economic success, political challenges, foreign policy crises, and personal scandals. Despite controversy, he completed his term with significant achievements and a lasting impact on American politics.

- His presidency was marked by strong economic growth with low unemployment and a budget surplus.
- His administration faced political challenges, including the 1994 Republican congressional sweep.
- He dealt with foreign policy crises in Bosnia, Kosovo, and the Middle East.
- Scandals, most notably the Monica Lewinsky affair, led to his 1998 impeachment.
- He was acquitted by the Senate and remained in office until the end of his second term.

Accomplishments

- Balanced the federal budget, achieving a surplus by 1998.
- Signed the North American Free Trade Agreement (NAFTA) in 1993.
- Passed the Family and Medical Leave Act in 1993.
- Oversaw the 1996 Welfare Reform Act, reshaping social welfare programs.
- Facilitated the Oslo Accords, advancing Israeli-Palestinian peace efforts.
- Expanded NATO and supported peacekeeping in Bosnia and Kosovo.

First Lady's Contributions

- Led the Task Force on National Health Care Reform, though the plan failed to pass.

- Advocated for women's and children's rights globally.
- Established the Children's Health Insurance Program (CHIP) in 1997.
- Promoted early childhood education and foster care reforms.
- Authored *It Takes a Village*, emphasizing community support for children.

Positive Traits

Clinton's natural charisma and strong communication skills helped him connect with diverse audiences, while his adaptability allowed him to succeed in a shifting political landscape.

- He was known for his charisma, empathy, and ability to connect with people.
- His communication skills made him an effective public speaker.
- His adaptability and political acumen helped him navigate legislative battles.
- He maintained public support even during times of controversy.

Negative Traits

Despite his strengths, Clinton faced challenges that undermined his credibility and effectiveness at times.

- His personal scandals, particularly the Lewinsky affair, damaged his reputation.

- These scandals distracted from his political agenda.
- Critics noted a tendency toward indecision.
- His reliance on polling sometimes created perceptions of opportunism.

Effects on Presidency

Clinton's strengths and weaknesses had lasting impacts on his leadership and legacy.

- His positive traits fostered bipartisan cooperation, including welfare reform and trade agreements.
- His scandals fueled partisan divides and led to his impeachment.
- His resilience allowed him to complete his term successfully.
- He left office with high approval ratings, though his legacy remains polarizing.

Pets

The Clintons shared their White House years with beloved pets who captured public attention.

- They had a chocolate Labrador named Buddy.
- They also had a cat named Socks, who became a White House celebrity.
- Socks was often featured in the media and public appearances.

Religious Persuasion

Clinton's faith was an important influence in both his personal and political life.

- He is a Southern Baptist.
- He attended church regularly throughout his life.
- He often referenced faith in his public speeches.
- His religious background influenced his emphasis on compassion and community in policy.

Interesting Anecdotes

Clinton's personality and cultural appeal were often highlighted through memorable public moments that helped shape his image.

- His saxophone performance on *The Arsenio Hall Show* in 1992 became an iconic campaign moment.
- The performance showcased his ability to connect with younger and more diverse audiences.
- He once jogged to a McDonald's while serving as governor of Arkansas.
- This lighthearted act reflected his relatable and down-to-earth persona.

Ages at Death

- Bill Clinton was born on August 19, 1946, and is 78 years old.
- Hillary Clinton was born on October 26, 1947, and is 77 years old.

- Both are currently alive.

Causes of Death

Not applicable, as both Bill and Hillary Clinton are living.

Burial Locations

Not applicable, as both are living.

Conclusion

Bill Clinton's presidency, defined by economic prosperity and personal controversies, reflects a complex legacy. His ability to connect with people and navigate political challenges earned him the "Comeback Kid" moniker, while his scandals left lasting divisions. Alongside Hillary's advocacy, his tenure shaped modern Democratic politics and global diplomacy.

George W. Bush: Architect of the Post-9/11 Era

Introduction

George Walker Bush, the 43rd President of the United States, served from 2001 to 2009, a period defined by transformative global and domestic challenges. Born into a prominent political family, Bush's life trajectory—from a Texas oilman to governor and then president—was shaped by ambition, faith, and a commitment to public service. His presidency is most remembered for its response to the September 11, 2001, terrorist attacks, which redirected U.S. foreign policy toward the global war on terror and led to wars in Afghanistan and Iraq. Bush's leadership style, marked by decisiveness and controversy, continues to spark debate about his legacy.

Description

George W. Bush, often referred to as "W" to distinguish him from his father, George H.W. Bush, was a Republican politician and businessman known for his folksy demeanor and steadfast resolve. His presidency navigated unprecedented national security threats and economic turmoil, earning him both ardent supporters and fierce critics.

Early Life

George W. Bush was born on July 6, 1946, in New Haven, Connecticut, but grew up in Midland and Houston, Texas, where his father pursued a career in the oil industry. His early years were marked by a privileged upbringing and a competitive family environment.

- Attended Kinkaid School and later Phillips Academy in Andover, Massachusetts (1961–1964), where he was a cheerleader and played baseball.

- Earned a bachelor's degree in history from Yale University in 1968, where he was a member of the Skull and Bones society.

- Served in the Texas Air National Guard (1968–1974) as an F-102 fighter pilot, avoiding combat during the Vietnam War.

- Received an MBA from Harvard Business School in 1975, becoming the first U.S. president with an MBA.

Family

Bush's family ties were deeply influential, rooted in a legacy of wealth and public service.

- Father: George Herbert Walker Bush, 41st U.S. President (1989–1993) and former vice president, CIA director, and congressman.
- Mother: Barbara Pierce Bush, a supportive matriarch known for her literacy advocacy as First Lady.
- Siblings: Five, including Jeb Bush (former Florida governor), Neil, Marvin, Dorothy, and Pauline ("Robin"), who died of leukemia in 1953 at age three, profoundly affecting the family.
- Married Laura Welch, a teacher and librarian, in 1977 after a three-month courtship.

Children

George and Laura Bush have two children, twin daughters born in 1981.

- Jenna Welch Bush Hager: A television personality and author, married to Henry Hager, with three children. She worked as a teacher and interned for UNICEF during her father's presidency.
- Barbara Pierce Bush: A global health advocate, co-founder of Global Health Corps, married to Craig Coyne, with one child. She maintained a lower public profile during her father's presidency.

Rise to Power

Bush's political ascent was marked by business ventures, personal transformation, and leveraging his family's network.

- Worked in the Texas oil industry, founding Arbusto Energy in 1977, which later merged with other firms.
- Co-owned the Texas Rangers baseball team (1989–1994), serving as managing general partner, boosting his public profile.
- Elected governor of Texas in 1994, defeating incumbent Ann Richards, and re-elected in 1998 with 68% of the vote.
- Ran for president in 2000, winning the Republican nomination and defeating Vice President Al Gore in a contested election decided by the Supreme Court in *Bush v. Gore*, securing 271 electoral votes despite losing the popular vote.

Influences

Bush's worldview was shaped by personal experiences, faith, and key figures.

- His father's political career inspired his public service ethos but also set high expectations.
- A spiritual awakening in the mid-1980s, guided by Rev. Billy Graham, deepened his Christian faith, influencing his "compassionate conservatism."
- Advisers like Karl Rove (political strategist) and Condoleezza Rice (national security adviser) shaped his campaign and foreign policy.
- The 9/11 attacks profoundly altered his presidency, prioritizing national security over domestic reform.

Party Affiliation

- Republican Party: Bush aligned with the GOP's conservative wing, advocating limited government, tax cuts, and strong national defense. His "compassionate conservatism" sought to blend traditional Republican values with social outreach, though some conservatives criticized his spending policies.

Presidency

Bush's presidency (2001–2009) was defined by the 9/11 attacks, which shifted his focus from domestic priorities like education reform to a global war on terror. His administration launched wars in Afghanistan (2001) and Iraq (2003), reshaped national security policy, and faced economic challenges culminating in the 2008 financial crisis.

- The 2000 election controversy and *Bush v. Gore* decision polarized the nation, complicating his early term.
- Post-9/11, Bush's approval ratings soared to 90%, but later fell due to prolonged wars and economic woes.
- Domestic policies included tax cuts (2001, 2003), No Child Left Behind, and Medicare expansion, while foreign policy emphasized preemption and democracy promotion.
- Hurricane Katrina (2005) response drew criticism for federal delays, damaging his public image.
- The 2008 financial crisis prompted controversial bailouts (TARP), leaving a mixed economic legacy.

Accomplishments

- Passed the No Child Left Behind Act (2001), raising education standards with bipartisan support, though criticized for over-testing.

- Signed the Medicare Modernization Act (2003), adding prescription drug benefits, the largest Medicare expansion since 1965.

- Implemented tax cuts (2001, 2003), reducing rates to stimulate the economy, though contributing to deficits.

- Launched the President's Emergency Plan for AIDS Relief (PEPFAR) in 2003, saving over 25 million lives by 2023, primarily in Africa.

- Created the Department of Homeland Security (2002) and passed the Patriot Act (2001) to enhance national security post-9/11.

- Overthrew Taliban in Afghanistan (2001) and Saddam Hussein in Iraq (2003), though long-term stability proved elusive.

- Established the President's Malaria Initiative (2005), reducing malaria deaths in 15 African countries.

First Lady's Contributions

Laura Bush, First Lady from 2001 to 2009, focused on education, literacy, and women's rights, earning praise for her grace under pressure.

- Promoted literacy through the National Book Festival, launched in 2001, now an annual event celebrating reading.

- Advocated for education, supporting teacher training and early childhood programs like Ready to Read, Ready to Learn.
- Championed women's rights globally, visiting Afghanistan to highlight women's education and health post-Taliban.
- Founded the Laura Bush Foundation for America's Libraries, providing grants to school libraries nationwide.
- Played a diplomatic role, delivering radio addresses and meeting world leaders to advance U.S. soft power.

Personal Positive Traits

- Decisiveness: Bush's swift response to 9/11, including military action in Afghanistan, rallied the nation and allies.
- Optimism: His upbeat demeanor fostered hope during crises, though sometimes perceived as disconnected from realities.
- Loyalty: Strong bonds with advisers like Dick Cheney and Donald Rumsfeld ensured a cohesive team, though it risked groupthink.
- Empathy: Private visits, like to Fort Hood shooting victims in 2009, showed personal warmth, enhancing his post-presidency image.

Personal Negative Traits

- Stubbornness: Resistance to adjusting the Iraq strategy early on, due to prolonged instability and eroding public support.

- Communication struggles: Malapropisms (e.g., "misunderestimated") and unclear messaging on policies fueled perceptions of incompetence.

- Over-reliance on advisers: Heavy dependence on Cheney and Rumsfeld led to accusations of being manipulated, particularly on Iraq.

- Perceived detachment: Elite background and phrases like "mission accomplished" (2003) alienated some Americans during economic hardship.

Effects on Presidency

Bush's traits shaped a presidency of bold action and deep division.

- Positive: Decisiveness and optimism unified the nation post-9/11, enabling the rapid passage of security measures and PEPFAR's success.

- Negative: Stubbornness and communication issues alienated moderates, while Iraq's mishandling and Katrina's fallout tanked approval ratings to 25% by 2008.

- His loyalty fostered stability but insulated him from dissent, delaying course corrections in foreign and domestic policy.

Pets

The Bush family's pets were beloved White House residents, reflecting their warmth.

- Barney: A Scottish Terrier, known for biting reporters, starred in White House Christmas videos.
- Miss Beazley: Another Scottish Terrier, Barney's companion, gifted to Laura in 2005.
- Spot Fetcher: An English Springer Spaniel, inherited from George H.W. Bush, euthanized in 2004 at age 15.
- India: A cat, named after a Texas Rangers player, lived quietly until her death in 2009.

Religious Persuasion

- Bush is a devout Methodist whose faith profoundly shaped his personal and political life.
- A 1986 recommitment to Christianity, spurred by Billy Graham, led him to quit drinking at age 40.
- Faith-informed policies like the Office of Faith-Based and Community Initiatives promote religious organizations' social work.
- Described himself as a "born-again Christian," emphasizing personal redemption and moral clarity in decision-making.

Interesting Anecdotes

- After 9/11, Bush threw the ceremonial first pitch at a 2001 World Series game in Yankee Stadium, wearing a bulletproof vest, symbolizing national resilience.

- During a 2008 trip to Iraq, an Iraqi journalist threw shoes at Bush, who dodged them with quick reflexes, later joking about the incident.
- As governor, Bush nicknamed staffers, like calling Karl Rove "Boy Genius," fostering a jovial but disciplined team dynamic.
- Post-presidency, Bush took up painting, creating portraits of veterans and world leaders, revealing a reflective side.

Ages at Death

- George W. Bush is alive as of June 20, 2025, aged 78.
- Laura Bush is alive, aged 78 (born November 4, 1946).

Causes of Death

- Not applicable, as both George W. Bush and Laura Bush are living.

Burial Locations

- Not applicable, as both are living. However, Bush has expressed a desire to be buried in Texas, likely near his presidential library at Southern Methodist University in Dallas.

Conclusion

George W. Bush's life and presidency reflect a complex legacy of resilience and controversy. From his Texas roots to the White House, his journey was shaped by family, faith, and the seismic events of 9/11, which defined his tenure as the architect of the post-9/11 era. His accomplishments, like PEPFAR and tax cuts, coexist with divisive decisions, notably the Iraq War and economic challenges. Laura Bush's literacy and global advocacy complemented his leadership. Bush's personal traits—decisiveness and optimism, tempered by stubbornness and communication struggles—drove both his successes and setbacks. Today, his post-presidency work, including painting and veteran support, underscores a commitment to service, leaving historians to debate his impact as a leader who

Barack Obama: The First African American President

Introduction

Barack Obama, the 44th President of the United States, served from January 20, 2009, to January 20, 2017. His presidency marked a historic milestone as the first African American to hold the office, symbolizing a shift in American racial and political dynamics. Known for his eloquence, progressive ideals, and efforts to bridge divides, Obama's leadership focused on healthcare reform, economic recovery, and global diplomacy. His tenure, while transformative for many, also faced significant challenges and polarization. This summary explores his life, presidency, and legacy, alongside the contributions of First Lady Michelle Obama.

Early Life

Barack Hussein Obama Jr. was born on August 4, 1961, in Honolulu, Hawaii. His early years were shaped by a diverse cultural background and a peripatetic upbringing.

- His mother, Ann Dunham, was a white American from Kansas, and his father, Barack Hussein Obama Sr., was a Black Kenyan who met Ann while studying at the University of Hawaii.
- His parents separated when he was two, and his father returned to Kenya, leaving Obama primarily raised by his mother and maternal grandparents in Hawaii.
- At age six, Obama moved with his mother and her new husband to Indonesia, where he lived for four years before returning to Hawaii.
- He attended Punahou School, a prestigious private institution in Honolulu, graduating in 1979.
- Obama later moved to Los Angeles to attend Occidental College before transferring to Columbia University in New York, earning a bachelor's degree in political science in 1983.
- His early experiences with racial identity and diverse environments influenced his worldview and later political philosophy.

Family

Obama's family life reflects a blend of cultural influences and strong personal bonds.

- His mother, Ann Dunham, was an anthropologist who instilled in him a sense of curiosity and social responsibility; she passed away in 1995.

- His father, Barack Hussein Obama Sr., was an economist who had limited contact with Obama after the divorce; he died in a car accident in 1982.

- Obama was close to his maternal grandparents, Madelyn and Stanley Dunham, who played a significant role in his upbringing in Hawaii.

- He married Michelle LaVaughn Robinson in 1992, forming a partnership that became a cornerstone of his personal and political life.

- The Obamas maintained a close-knit family, prioritizing stability for their children despite the demands of public life.

Children

Barack and Michelle Obama have two daughters, who were raised in the public eye during their White House years.

- Malia Ann Obama, born July 4, 1998, pursued studies at Harvard University and has explored a career in filmmaking.

- Natasha Marian Obama, known as Sasha, born June 10, 2001, attended the University of Michigan, and has kept a lower public profile.

- The Obamas emphasized normalcy for their daughters, ensuring they attended school, did chores, and maintained privacy despite their high-profile status.

Rise to Power

Obama's ascent to the presidency was marked by rapid political success and a knack for inspiring diverse audiences.

- After Columbia, he worked as a community organizer in Chicago's South Side from 1985 to 1988, focusing on grassroots activism and social justice.
- He attended Harvard Law School, earning a law degree in 1991 and becoming the first African American president of the Harvard Law Review.
- Obama taught constitutional law at the University of Chicago and served as an Illinois State Senator from 1997 to 2004, championing healthcare and ethics reforms.
- His 2004 Democratic National Convention keynote address, emphasizing unity, catapulted him to national prominence.
- Elected to the U.S. Senate in 2004, he served until 2008, when he announced his presidential candidacy.
- In 2008, Obama won the Democratic nomination over Hillary Clinton and defeated Republican John McCain, capitalizing on a message of hope and change.

Influences

Obama's worldview was shaped by a range of intellectual, cultural, and personal influences.

- His mother's progressive ideals and commitment to education instilled a sense of purpose and global awareness.

- Community organizing in Chicago exposed him to the challenges of marginalized communities, grounding his policy priorities.

- Figures like Martin Luther King Jr. and Abraham Lincoln inspired his focus on equality and unity.

- His Christian faith, developed later in life, influenced his moral framework and public rhetoric.

- Legal scholars and constitutional law shaped his approach to governance and judicial philosophy.

Party Affiliation

- Obama is a member of the Democratic Party.

- He aligned with the party's progressive wing, advocating for social equity, healthcare reform, and environmental protection.

- His pragmatic approach often sought bipartisan solutions, though he faced significant Republican opposition during his presidency.

Presidency

Obama's presidency (2009–2017) navigated major domestic and global challenges, including the Great Recession, healthcare reform, and international conflicts.

- He assumed office during an economic crisis, signing the American Recovery and Reinvestment Act in 2009 to stimulate the economy.

- The Affordable Care Act (ACA), or "Obamacare," became his signature legislative achievement, expanding healthcare access.

- He oversaw the withdrawal of U.S. troops from Iraq by 2011 and authorized the operation that killed Osama bin Laden in 2011.

- His administration advanced climate initiatives, including the Paris Climate Agreement in 2015.

- Obama faced criticism for drone warfare, slow economic recovery, and partisan gridlock, particularly with a Republican-controlled Congress after 2010.

- His foreign policy emphasized diplomacy, including the Iran nuclear deal, though tensions persisted in the Middle East and with Russia.

Accomplishments

- Passed the Affordable Care Act, providing health insurance to millions of uninsured Americans.

- Signed the Dodd-Frank Act to regulate Wall Street and prevent future financial crises.

- Authorized the mission that killed Osama bin Laden, a major blow to al-Qaeda.

- Advanced marriage equality, with his administration supporting the 2015 Supreme Court decision legalizing same-sex marriage.

- Negotiated the Paris Climate Agreement to address global climate change.
- Restored diplomatic relations with Cuba in 2014, ending decades of estrangement.
- Reduced the U.S. prison population through clemency and sentencing reforms.

First Lady's Contributions

- Launched the Let's Move! Campaign to combat childhood obesity through healthy eating and exercise.
- Initiated Joining Forces to support military families with education and job opportunities.
- Promoted girls' education globally through the Let Girls Learn initiative.
- Advocated for healthy school lunches, influencing national nutrition standards.
- Published her memoir, *Becoming*, in 2018, inspiring millions with her story of resilience.

Positive Traits and Effects

Obama's personal strengths significantly shaped his presidency.

- His charisma and oratorical skills inspired a broad coalition of voters, fostering optimism during his 2008 campaign.
- A calm, deliberative demeanor helped him navigate crises like the 2008 financial collapse and foreign policy challenges.

- His intellectual rigor, honed by his legal background, informed complex policy decisions like the ACA.
- Empathy, drawn from his diverse upbringing, connected him with marginalized groups and informed social justice policies.
- These traits bolstered his global image and domestic support, though some critics argued his idealism hindered pragmatic deal-making.

Negative Traits and Effects

Obama's weaknesses also influenced his tenure.

- His preference for consensus sometimes led to compromises that frustrated progressive allies, such as on healthcare reform.
- A perceived aloofness occasionally strained relationships with congressional leaders, limiting legislative wins.
- His reliance on executive actions, especially later in his presidency, drew accusations of overreach from opponents.
- Critics noted a cautious approach to foreign policy, which some argued emboldened adversaries like Russia in Syria and Ukraine.
- These traits contributed to partisan gridlock and limited the scope of his domestic agenda.

Pets

The Obama family had two pets during their White House years.

- Bo, a Portuguese Water Dog, joined the family in 2009 as a gift from Senator Ted Kennedy.

- Sunny, another Portuguese Water Dog, arrived in 2013, becoming Bo's companion.

- Both dogs were beloved by the family and became public symbols of warmth and relatability.

Religious Persuasion

- Obama identifies as a Christian, having joined the United Church of Christ in the 1980s while working in Chicago.

- His faith, influenced by progressive Christian values, emphasized social justice and community service.

- He attended Trinity United Church of Christ in Chicago, though controversies involving its pastor, Jeremiah Wright, sparked debate during the 2008 campaign.

- In the White House, Obama did not regularly attend a specific church but participated in private prayer and occasional services.

Interesting Anecdotes

- During his 2008 campaign, Obama's speech at the Democratic National Convention was so impactful that it reportedly moved Oprah Winfrey to tears, boosting his national profile.

- He once played basketball with NBA stars like LeBron James to raise funds for wounded veterans, showcasing his love for the sport.

- Obama's habit of sneaking cigarettes during his early political career became a personal challenge; he quit smoking before his presidency.

- He and Michelle famously crashed a White House tour group to greet surprised visitors, a rare moment of spontaneity.

Ages at Death, Causes of Death, and Burial Locations

- As of June 20, 2025, Barack Obama (born August 4, 1961) is 64 years old and alive.

- Michelle Obama (born January 17, 1964) is 61 years old and alive.

- No information on causes of death or burial locations is available, as both are living.

Conclusion

Barack Obama's life and presidency represent a transformative chapter in American history. As the first African American president, he broke barriers and inspired a generation with his message of hope and unity. His legislative achievements, particularly the Affordable Care Act, reshaped domestic policy, while his global diplomacy sought to restore America's international standing. Michelle Obama's advocacy for health, education, and military families complemented his legacy. Despite challenges, including partisan divides and criticism of his leadership style, Obama's intellect, empathy, and resilience left an indelible mark. His story, rooted in a diverse upbringing and driven by a commitment to progress, continues to influence American politics and society.

Donald John Trump: Architect of MAGA and Polarizing Political Figure

Introduction

Donald John Trump, born June 14, 1946, in Queens, New York, is an American politician, businessman, and media personality who served as the 45th president of the United States (2017–2021) and the 47th president (2025–present). A polarizing figure, Trump reshaped American politics with his "Make America Great Again" (MAGA) movement, blending right-wing populism with a combative, outsider persona. His career spans real estate mogul, reality TV star, and political leader, marked by unprecedented controversy, legal battles, and a loyal base. He is most remembered for his MAGA movement, which redefined the Republican Party and fueled intense national division.

Early Life

Donald Trump was born into wealth, the fourth of five children in a family rooted in real estate. His upbringing in

Queens and education shaped his brash, ambitious character.

- Grew up in a mansion in Jamaica Estates, Queens, with strict parental oversight.
- Attended Kew-Forest School until age 13, when behavioral issues led his parents to enroll him in New York Military Academy (1959–1964).
- Excelled socially and athletically at the academy, enjoying its disciplined environment.
- Studied at Fordham University (1964–1966) before transferring to the University of Pennsylvania's Wharton School, earning a bachelor's degree in economics in 1968.
- Avoided the Vietnam War draft through college and medical deferments (bone spurs diagnosis).
- By age eight, he was a millionaire in inflation-adjusted dollars, receiving $20,000 annually from his father's business.

Family

Trump's family background, tied to his father's real estate empire and his mother's Scottish heritage, instilled a drive for success and public recognition.

- Father, Fred Trump, a successful real estate developer, provided financial and political connections.
- Mother, Mary Anne MacLeod Trump, a Scottish immigrant, emphasized religious education.

- Siblings included Maryanne (retired judge), Fred Jr. (deceased pilot), Elizabeth (retired banker), and Robert (deceased businessman).
- Fred Jr.'s alcoholism and early death at 43 profoundly influenced Trump's decision to abstain from alcohol, cigarettes, and drugs.
- Family wealth, estimated at $250–300 million upon Fred's death in 1999, shaped Trump's privileged lifestyle.

Children

Trump has five children from three marriages, several of whom have played significant roles in his business and political endeavors.

- Donald Jr. (born 1977), Ivanka (born 1981), and Eric (born 1983) from the first marriage to Ivana Zelníčková (1977–1992).
- Tiffany (born 1993) from the second marriage to Marla Maples (1993–1999).
- Barron (born 2006) from the third marriage to Melania Knauss (2005–present).
- Donald Jr., Ivanka, and Eric managed the Trump Organization during his first presidency; Ivanka and her husband, Jared Kushner, served as senior White House advisors.
- Barron, largely private, and Kai, Donald Jr.'s daughter, emerged politically, with Kai speaking at the 2024 Republican National Convention.

Rise to Power

Trump's ascent from real estate tycoon to political juggernaut was marked by media savvy, business ventures, and a knack for controversy.

- Took over his father's real estate business in 1971, renaming it the Trump Organization.
- Built high-profile projects like the Grand Hyatt (1980) and Trump Tower (1983), leveraging tax abatements and his father's political ties.
- Expanded into casinos, hotels, and golf courses, though faced six business bankruptcies in the 1990s and 2000s.
- Gained fame as host of *The Apprentice* (2004–2015), cementing his image as a billionaire dealmaker.
- Entered politics in 2015, capitalizing on populist rhetoric and media attention to win the 2016 Republican nomination and presidency against Hillary Clinton.
- Survived two assassination attempts in 2024, bolstering his image as a resilient figure.

Influences

Trump's worldview was shaped by family, business mentors, and cultural figures, fostering his aggressive, self-promotional style.

- Father Fred Trump instilled a relentless work ethic and strategic use of political connections.
- Attorney Roy Cohn, a ruthless litigator, taught Trump to counterattack and never admit fault.

- Norman Vincent Peale, a pastor promoting positive thinking, influenced Trump's optimism and self-belief.
- Media exposure through *The Apprentice* and tabloid coverage shaped his understanding of public perception.
- Political influences included populist and nationalist movements, aligning with his "America First" stance.

Party Affiliation

Trump's political journey reflects shifts in allegiance, ultimately transforming the Republican Party.

- Initially a Republican, he switched to the Reform Party in 1999, briefly exploring a 2000 presidential run.
- Registered as a Democrat (2001–2009) before returning to the Republican Party in 2009.
- Ran as a Republican in 2016 and 2024, reshaping the party with MAGA ideology.
- Embraced right-wing populism, prioritizing immigration control, trade protectionism, and distrust of elites.
- His influence sidelined moderate Republicans, making the party more aligned with his base.

Presidency

Trump's presidencies (2017–2021, 2025–present) were defined by bold policies, controversy, and historic legal challenges.

- **First Term (2017–2021):** Focused on tax reform, deregulation, and conservative judicial appointments.

- **Second Term (2025–present):** Began with aggressive executive actions on immigration, trade, and government efficiency.

- Impeached twice: in 2019 for abuse of power and obstruction (Ukraine scandal) and in 2021 for incitement of insurrection (January 6 Capitol riot); acquitted both times by the Senate.

- Convicted in 2024 of 34 felony counts of falsifying business records, receiving an unconditional discharge in 2025.

- His refusal to concede the 2020 election to Joe Biden and claims of voter fraud fueled division and the Capitol riot.

- Survived two assassination attempts in 2024, shaping his second term's narrative of defiance.

Accomplishments

- Signed the Tax Cuts and Jobs Act of 2017, reducing corporate and personal income taxes.

- Appointed three Supreme Court justices (Gorsuch, Kavanaugh, Barrett), cementing a conservative majority.

- Launched Operation Warp Speed, accelerating COVID-19 vaccine development.

- Renegotiated trade deals, including USMCA (replacing NAFTA) and agreements with China, Japan, and South Korea.

- Implemented aggressive immigration policies, including border wall construction and travel bans.

- Reduced federal regulations, removing 25,000 pages from the Federal Register.
- Established the White House Faith and Opportunity Initiative to protect religious liberty.
- Defeated ISIS's territorial caliphate, strengthening the U.S. military presence in the Middle East.
- Modernized the National Environmental Policy Act, speeding up infrastructure project approvals.
- Survived two assassination attempts in 2024, reinforcing his political resilience.

First Lady's Contributions

Melania Trump, First Lady during both terms, focused on children's issues and public image management.

- Launched the Be Best campaign, promoting children's well-being, online safety, and opioid crisis awareness.
- Oversaw White House restoration projects, preserving historical decor and artifacts.
- Advocated for foster care and adoption reforms to support faith-based organizations.
- Represented the U.S. internationally, promoting American values and diplomacy.
- Published a memoir in 2024, sharing her experiences and perspectives as First Lady.

Personal Positive Traits

Trump's charisma, media savvy, and resilience have been central to his political success, though they have also polarized.

- Charismatic communicator, adept at rallying supporters with direct, unfiltered rhetoric.
- Resilient under pressure, as seen in surviving legal battles and assassination attempts.
- Media mastery, leveraging *The Apprentice* and social media (notably Twitter) to shape public perception.
- Strategic optimism, drawing from Norman Vincent Peale's teachings, projects confidence.
- Business acumen, building a global brand despite financial setbacks.
- These traits energized his base, disrupted traditional politics, and maintained his influence post-2020.

Personal Negative Traits

Trump's combative style and controversial rhetoric often deepened national divides.

- Propensity for false or misleading statements, promoting conspiracy theories like birtherism.
- Authoritarian tendencies include targeting political opponents and challenging democratic norms.
- Divisive rhetoric, disparaging minorities, and inflaming racial tensions, which scholars link to his support base.

- Lack of traditional political experience leads to chaotic governance and ethical concerns.
- Refusal to concede the 2020 election, culminating in the January 6 Capitol riot.
- These traits eroded trust in institutions, alienated allies, and fueled legal and social backlash.

Effects on Presidency

Trump's traits profoundly shaped his presidencies, driving both achievements and controversies.

- Positive traits enabled bold policy moves, like tax reform and judicial appointments, by mobilizing Republican support.
- His media dominance kept him in the spotlight, amplifying his agenda but also controversies.
- Negative traits led to historic impeachments, legal challenges (over 300 lawsuits by May 2017), and international disapproval (U.S. leadership ratings dropped to 16% globally).
- His refusal to concede in 2020 undermined democratic norms, leading to the Capitol riot and a second impeachment.
- In his second term, fewer ethical guardrails and a loyal administration suggest a more aggressive agenda, potentially deepening polarization.

Pets

The Trump family has not been publicly associated with pets during his presidency, a departure from many White House traditions.

- No known pets resided in the White House during Trump's terms.
- Trump has expressed disinterest in having pets, reportedly calling dog ownership "phony."
- This absence contrasts with predecessors like the Obamas (dogs Bo and Sunny) and reflects his focus on personal branding over traditional presidential norms.

Religious Persuasion

Trump's religious affiliation has been a point of contention, often seen as more performative than deeply held.

- Identified as a Presbyterian Protestant in 2016, later claiming nondenominational Christian status in 2020.
- A 2017–2021 survey found 63% of Americans doubted his religiosity, with 44% believing he was Christian.
- Critics note superficial knowledge of Christianity, with biographers describing him as not particularly religious.
- Supported religious liberty, attended the March for Life, and established the White House Faith and Opportunity Initiative.
- His appeals to evangelical voters were strategic, aligning with conservative social policies.

Interesting Anecdotes

Trump's life is filled with colorful stories reflecting his personality and public image.

- In 2007, he participated in WWE's "Battle of the Billionaires" at WrestleMania 23, shaving Vince McMahon's head in a staged event that set pay-per-view records.
- During his 2025 term, he hung a portrait of himself from the July 2024 assassination attempt in the White House, overshadowing other presidential portraits.
- Rebuilt New York's Wollman Skating Rink in 1986 in four months for $1.8 million, after the city's seven-year, $20 million failure, boosting his reputation as a problem-solver.
- His 2015 campaign announcement, descending a Trump Tower escalator, became an iconic moment of his political rise.
- Hosted *Saturday Night Live* in 2004 and 2015, achieving high ratings and reinforcing his media presence.

Ages at Death

As of June 20, 2025, both Donald Trump and Melania Trump are alive.

- Donald Trump is 79 years old (born June 14, 1946).
- Melania Trump is 55 years old (born April 26, 1970).
- No death-related information applies at this time.

Causes of Death

- Not applicable, as both Donald Trump and Melania Trump are living as of June 20, 2025.

Burial Locations

- Not applicable, as both Donald Trump and Melania Trump are alive as of June 20, 2025.

Conclusion

Donald Trump's life and presidencies are a study in contrasts: a billionaire outsider who disrupted American politics with MAGA, achieving significant policy wins while sparking historic controversy. His charisma and resilience drove landmark reforms, but his divisive rhetoric and norm-breaking actions deepened polarization, legal challenges, and global skepticism. Melania Trump's reserved First Lady role complemented his style, focusing on children's issues and White House preservation. With no pets and a questionable religious commitment, Trump's legacy is defined by his transformation of the Republican Party and unwavering base loyalty, balanced against accusations of authoritarianism and democratic erosion. As his second term unfolds, Trump's impact—bold, polarizing, and unprecedented—continues to shape America's political landscape.

Joe Biden: Champion of Resilience and Unity

Introduction

Joseph Robinette Biden Jr., the 46th President of the United States, served from January 20, 2021, to January 20, 2025, and is most remembered for his resilience in the face of personal tragedy and his commitment to restoring unity in a polarized nation. A career politician with decades of experience, Biden's journey from a working-class upbringing in Scranton, Pennsylvania, to the White House was marked by perseverance, empathy, and a dedication to public service. His presidency navigated unprecedented challenges, including a global pandemic, economic recovery, and international conflicts, while his personal life was shaped by profound loss and a deep-rooted Catholic faith. This summary explores Biden's life, family, political career, and legacy, alongside the contributions of First Lady Jill Biden and the complexities of his presidency.

Early Life

Joe Biden was born on November 20, 1942, in Scranton, Pennsylvania, to Catherine Eugenia "Jean" Finnegan Biden and Joseph Robinette Biden Sr. Growing up in a working-class Irish Catholic family, his early years were shaped by financial struggles and personal challenges.

- His father, once affluent during World War II, faced financial setbacks, forcing the family to live with Biden's maternal grandparents for several years.

- In 1953, the family relocated to Claymont, Delaware, and later to Mayfield, Delaware, where Biden's father worked as a used-car salesman, stabilizing the family in a middle-class lifestyle.

- Biden attended Archmere Academy, where he excelled as class president in his junior and senior years and played football, overcoming a severe stutter that toughened his resolve.

- He graduated from the University of Delaware in 1965 with a double major in history and political science and earned a Juris Doctor from Syracuse University College of Law in 1968, ranking 76th in his class of 85.

- Initially identifying with Republican ideals, Biden registered as an Independent due to his disapproval of Richard Nixon, later aligning with the Democratic Party.

Family

Biden's family life was marked by love, loss, and resilience, with two marriages and a close-knit extended family.

- He married Neilia Hunter in 1966, a student at Syracuse University, and they had three children: Beau, Hunter, and Naomi.
- In 1972, tragedy struck when Neilia and Naomi died in a car accident, leaving Biden to raise his two young sons, Beau and Hunter, as a single father.
- In 1975, Biden met Jill Tracy Jacobs on a blind date, and they married in 1977 at the United Nations chapel in New York City, honeymooning in Hungary's Lake Balaton.
- Jill, an educator with a Ph.D., became a stabilizing force, helping raise Beau and Hunter and giving birth to their daughter, Ashley, in 1981.
- Biden's family, of Irish and English descent, remained central to his life, with his sister Valerie managing his campaigns and his brothers, Francis and James, supporting his endeavors.

Children

Joe Biden fathered four children across his two marriages, with two surviving into adulthood.

- Joseph "Beau" Biden III (1969–2015) served as Delaware's Attorney General and was a major in the Delaware Army National Guard; he died of brain cancer at age 46, profoundly impacting Biden's emotional life.
- Robert Hunter Biden (born 1970) worked as a lobbyist and investment adviser, facing significant scrutiny for his business dealings, particularly with Ukrainian company Burisma, and legal troubles, including a 2024

federal tax evasion guilty plea, later pardoned by his father.

- Naomi Christina "Amy" Biden (1971–1972) died at 13 months in the 1972 car accident alongside her mother.
- Ashley Blazer Biden (born 1981), a social worker, activist, and fashion designer, has maintained a lower public profile but contributed to social justice initiatives.
- Biden's seven grandchildren, including five from Hunter and two from Beau, made him the first U.S. president to become a great-grandfather during his term.

Rise to Power

Biden's political career spanned over five decades, marked by tenacity and strategic alliances.

- In 1970, he won a seat on the New Castle County Council, followed by an upset victory in 1972 against incumbent Republican Senator J. Caleb Boggs, becoming one of the youngest U.S. senators at age 30.
- He served as a U.S. Senator from Delaware from 1973 to 2009, winning reelection six times and becoming the state's longest-serving senator.
- As a senator, Biden chaired the Senate Judiciary Committee and Foreign Relations Committee, overseeing contentious Supreme Court confirmation hearings for Robert Bork and Clarence Thomas.
- His 1988 presidential campaign ended due to plagiarism allegations involving British Labour leader

Neil Kinnock's speeches, and his 2008 bid faltered after a fifth-place finish in the Iowa caucus.

- Selected as Barack Obama's vice presidential running mate in 2008, Biden served two terms as the 47th Vice President (2009–2017), leveraging his Senate experience to advance Obama's agenda.

- In 2020, Biden won the Democratic presidential nomination, defeating Donald Trump with 306 electoral votes to 232, becoming the oldest president at inauguration at age 78.

Influences

Biden's worldview and leadership were shaped by personal tragedy, faith, and political mentors.

- The 1972 car accident that killed his wife and daughter instilled a deep empathy for loss, influencing his "comforter-in-chief" approach during the COVID-19 pandemic.

- His Catholic faith, tested by personal tragedies, reinforced his belief in human goodness and community, shaping his humanistic policy agenda.

- John F. Kennedy's idealism inspired Biden's political ambitions, while his Senate mentors, including Hubert Humphrey, influenced his focus on civil rights and social justice.

- His son Beau's life and death were pivotal, with Beau serving as an emotional anchor and inspiring Biden's Cancer Moonshot initiative.

- Jill Biden's resilience and educational advocacy encouraged Biden's renewed focus on politics and family after his early losses.

Party Affiliation

- Biden has been a lifelong member of the Democratic Party since registering as a Democrat in the late 1960s, after briefly considering himself a Republican and registering as an Independent.

- His moderate Democratic stance emphasized bipartisanship, though he faced criticism from progressives for policies like the 1994 Crime Bill.

- As president, he navigated a polarized Democratic Party, balancing progressive demands with centrist pragmatism, evident in his support for the Inflation Reduction Act and infrastructure investments.

Presidency

Biden's presidency (2021–2025) was defined by efforts to address the COVID-19 pandemic, economic recovery, and global leadership amid domestic polarization.

- He took office during a global health crisis, economic downturn, and the aftermath of the January 6 Capitol riot, aiming to "restore the soul of America."

- His administration faced challenges like high inflation (peaking at 9.1% in 2022), the Afghanistan withdrawal, and allegations of mental decline, culminating in his withdrawal from the 2024 presidential race after a weak debate performance.

- Biden endorsed Vice President Kamala Harris, who lost to Donald Trump, making Biden the second president to be succeeded by his predecessor.

- His approval ratings fell from over 50% in 2021 to 39% by December 2024, reflecting public frustration with inflation and foreign policy.

- Allegations of influence peddling centered on son Hunter's business dealings with Burisma, investigated by a House committee in 2023, though no charges were filed against Biden himself.

- Concerns about Biden's mental state grew after a June 2024 debate, with reports suggesting staff and family, including Jill, insulated him to protect his image.

- Questions about leadership arose, with critics alleging that unelected aides or Jill wielded undue influence, though Biden remained the decision-maker.

- Biden's use of an autopen for signing documents, including pardons, sparked debate about its legality and appropriateness, particularly for Hunter's 2024 pardon.

Accomplishments

- Signed the American Rescue Plan (2021), providing economic relief and boosting vaccine distribution during the COVID-19 crisis.

- Passed the Bipartisan Infrastructure Law (2021), investing in roads, bridges, and broadband access.

- Enacted the Inflation Reduction Act (2022), the largest climate investment in U.S. history, capping insulin costs at $35 for seniors.

- Signed the CHIPS and Science Act (2022), boosting domestic semiconductor production.

- Nominated Ketanji Brown Jackson as the first Black woman on the Supreme Court.

- Passed the Respect for Marriage Act (2022), protecting same-sex marriage rights.

- Oversaw a record 16 million new jobs, the strongest economic recovery among G7 nations post-COVID.

- Strengthened NATO and rallied global support for Ukraine against Russia's 2022 invasion.

- Launched the Cancer Moonshot, investing $1.8 billion to advance cancer research, and renamed it in honor of Beau Biden.

- Signed the PACT Act (2022), expanding benefits for veterans exposed to toxic substances.

First Lady's Contributions

- Championed the Joining Forces initiative, supporting military families through employment and education programs.

- Launched the White House Workforce Hub, fostering partnerships for career-connected learning.

- Promoted career-connected high school programs, securing grants to link education to job opportunities.

- Advocated for cancer patient navigation services, securing insurance reimbursements for 150 million Americans.
- Supported veterans' mental health, contributing to the National Maternal Mental Health Hotline and expanded VA services.

Personal Positive Traits

- Empathy: Biden's experiences with loss made him a compassionate leader, evident in his outreach to grieving families during the COVID-19 pandemic.
- Resilience: Overcame personal tragedies and political setbacks, maintaining a 50-year career through determination.
- Bipartisanship: His Senate experience fostered a willingness to work across party lines, as seen in the Bipartisan Infrastructure Law.
- Authenticity: His working-class roots and relatable demeanor earned him the "Middle-Class Joe" moniker, resonating with voters.

Personal Negative Traits

- Gaffes: Biden's propensity for verbal missteps, like his 1988 plagiarism scandal, fueled perceptions of carelessness.
- Stubbornness: His initial refusal to exit the 2024 race despite party pressure damaged his legacy.

- Perceived Frailty: Age-related concerns about his mental and physical health undermined public confidence, especially post-2024 debate.

- Family Loyalty: His defense of Hunter, including the 2024 pardon, raised ethical questions about favoritism.

Effects on Presidency

- Biden's empathy shaped his progressive agenda, like the American Rescue Plan, but his gaffes and perceived frailty amplified concerns about his leadership capacity.

- His stubbornness prolonged his 2024 campaign, alienating some Democrats and overshadowing his accomplishments.

- Family loyalty, particularly the Hunter pardon, fueled influence-peddling allegations, eroding public trust.

- Jill's protective role, while stabilizing, contributed to perceptions of an insulated administration, raising questions about who held influence behind the scenes.

Pets

- Biden had two German Shepherds, Champ and Major, during his vice presidency; Champ died in 2021, and Major was rehomed after biting incidents at the White House.

- In 2021, Biden received Commander, another German Shepherd, as a birthday gift from his brother and sister-in-law.

- The pets humanized Biden's public image but drew attention due to Major's behavioral issues, requiring rehoming.

Religious Persuasion

- Biden is a devout Roman Catholic, attending Mass regularly at St. Joseph's on the Brandywine in Greenville, Delaware.

- His faith was tested by the 1972 accident and Beau's death, leading to a belief in a more indifferent God but strengthening his commitment to Catholicism's communal aspects.

- His religious values informed policies like the Respect for Marriage Act and support for veterans, reflecting a humanistic outlook.

Interesting Anecdotes

- As a child, Biden climbed a 200-foot culm mountain in Scranton on a $5 dare, showcasing his daring nature despite the danger of collapsing ash pockets.

- During his Senate years, Biden commuted daily by Amtrak from Wilmington to Washington, D.C., to be with his sons, a practice he continued for 36 years.

- In 1988, Biden survived a life-threatening brain aneurysm, undergoing two surgeries and a seven-month Senate leave, demonstrating his physical resilience.

- He cited Chariots of Fire as his favorite film, admiring its theme of prioritizing principles over fame, reflecting his own values.

Ages at Death, Causes of Death, and Burial Locations

- As of July 1, 2025, Joe Biden is alive at age 82, diagnosed with aggressive prostate cancer that has metastasized to the bone, with treatment ongoing.
- Jill Biden, born June 3, 1951, is alive at age 74, with no reported serious health issues.
- No burial locations are applicable as both are living, and Biden has expressed a desire to spend retirement in Delaware, supporting the Beau Biden Foundation.

Conclusion

Joe Biden's life and presidency reflect a remarkable journey of resilience, shaped by personal tragedy, faith, and a commitment to public service. From his Scranton roots to the White House, he navigated challenges with empathy and determination, leaving a legacy of significant legislative achievements like the Inflation Reduction Act and the PACT Act. However, his presidency was marred by controversies over influence peddling, mental health concerns, and the use of an autopen, compounded by perceptions of Jill Biden's protective influence. Despite these, Biden's focus on unity and economic recovery, alongside Jill's advocacy for education and military families, cemented their impact. As he faces health challenges in retirement, Biden's story remains one of perseverance, faith, and a belief in the goodness of people, shaping a complex but enduring legacy.

Donald J. Trump: The Populist President

Description

Donald John Trump, born June 14, 1946, in Queens, New York, is an American politician, businessman, and media personality who served as the 45th president of the United States from 2017 to 2021 and returned as the 47th president in 2025. Known for his unorthodox political style, he reshaped American politics with a populist approach, leveraging his fame as a real estate mogul and reality TV star. His presidency was marked by polarizing policies, media controversies, and a loyal base galvanized by his "Make America Great Again" slogan.

Introduction

Trump's life is a story of ambition, reinvention, and controversy. From a privileged upbringing in New York to leading a global real estate empire, he transitioned into a political figure who defied traditional norms. His first term as president was defined by economic reforms, foreign policy shifts, and unprecedented legal challenges, including two impeachments. After losing in 2020, he won a second non-consecutive term in 2024, becoming only the second president in U.S. history to do so. His influence on American politics, often described as Trumpism, continues to spark debate about democracy, leadership, and national identity.

Early Life

Donald Trump was born into wealth as the fourth of five children in a family rooted in real estate. His upbringing, marked by competitiveness and discipline, shaped his early path toward business and leadership.

- He was born into a wealthy family with strong ties to real estate.
- His father, Fred Trump, a successful developer, strongly influenced his worldview.
- His youth was marked by competitiveness and behavioral challenges.
- At age 13, he was enrolled at the New York Military Academy, where he thrived in a structured environment.
- He excelled in athletics and leadership roles, though classmates described him as charismatic but prone to bending rules.
- He first attended Fordham University before transferring to the Wharton School at the University of Pennsylvania.
- He graduated in 1968 with a degree in economics.
- He avoided the Vietnam War draft through deferments, including a medical exemption for bone spurs.

- After college, he joined his father's real estate business, setting the stage for his future ventures.

Family

Trump's family life reflects both a foundation of wealth and a legacy of ambition. His parents' influence, combined with personal family challenges, shaped his values and decisions.

- His parents, Fred and Mary Anne MacLeod Trump, provided a base of wealth and opportunity.
- Fred Trump's real estate success and strict parenting fueled Donald Trump's drive.
- His mother, Mary, a Scottish immigrant, instilled a sense of family loyalty.
- He has siblings named Maryanne, Elizabeth, and Robert, while his brother Fred Jr. died at age 43.
- Fred Jr.'s struggles with alcoholism deeply affected Trump and led him to abstain from alcohol.
- His family's real estate legacy continues to define his public and private identity.

Children

Trump has five children from three marriages, several of whom have played significant roles in his business and political life.

- **Donald Trump Jr. (b. 1977)**: Eldest son of Ivana Zelníčková, a key figure in the Trump Organization and a vocal supporter during his father's campaigns.

- **Ivanka Trump (b. 1981)**: Daughter of Ivana, served as a senior White House advisor during Trump's first term, focusing on economic initiatives and women's empowerment.

- **Eric Trump (b. 1984)**: Son of Ivana, manages Trump Organization operations alongside his brother.

- **Tiffany Trump (b. 1993)**: Daughter of Marla Maples, less involved in politics but active in family branding.

- **Barron Trump (b. 2006)**: Son of Melania Knauss, largely kept out of the public eye during his father's political career.

Rise to Power

Trump's ascent began in the 1970s when he took over his father's company, renaming it the Trump Organization. He shifted focus to high-profile Manhattan projects, such as the Grand Hyatt and Trump Tower, earning a reputation for bold deals and media savvy. Despite financial setbacks, including six business bankruptcies in the 1990s and 2000s, Trump rebounded through branding ventures and reality TV fame with *The Apprentice* (2004–2015). His political journey started with a brief 2000 presidential run as a Reform Party candidate, but his 2015 Republican campaign, fueled by populist rhetoric and media dominance, led to a surprise 2016 victory over Hillary Clinton, despite losing the popular vote.

Influences

Trump's worldview was shaped by several key figures and philosophies:

- His father, Fred Trump, instilled a relentless work ethic and deal-making mentality.
- Norman Vincent Peale, a pastor whose book *The Power of Positive Thinking* influenced Trump's optimism and self-promotion.
- Roy Cohn, a ruthless attorney, mentored Trump in aggressive business and legal tactics.
- Political strategist Roger Stone and media figures like Steve Bannon amplified his populist messaging.
- Trump's exposure to New York's competitive real estate world and tabloid culture honed his knack for publicity and controversy.

Party Affiliation

Trump's political journey reflects ideological flexibility:

- **1987–1999**: Registered as a Republican, aligning with business-friendly policies.
- **1999–2001**: Switched to the Reform Party, exploring a presidential run.
- **2001–2009**: Registered as a Democrat, reflecting shifting views during this period.
- **2009–2011**: Returned to the Republican Party, embracing conservative rhetoric.
- **2011–2012**: Briefly independent, then rejoined Republicans for his 2016 campaign.
  His consistent Republican alignment since 2015

reflects a populist, anti-establishment stance, redefining the party's base through Trumpism.

Presidency

Trump's first term (2017–2021) was marked by bold policy moves, controversies, and two impeachments. His second term, beginning in 2025, continues his focus on economic nationalism and deregulation. Key aspects include:

- **First Term (2017–2021):** Emphasized tax cuts, deregulation, and conservative judicial appointments. Foreign policy included withdrawing from the Paris Climate Agreement and the Iran nuclear deal, moving the U.S. Embassy to Jerusalem, and brokering the Abraham Accords. His response to COVID-19, including Operation Warp Speed, was polarizing, with critics citing delayed action. Impeached in 2019 for pressuring Ukraine to investigate Joe Biden and in 2021 for inciting the January 6 Capitol riot, he was acquitted both times.

- **Second Term (2025–present):** Focused on tariffs, immigration enforcement, and dismantling federal agencies like the Department of Education. His administration faced legal challenges over executive actions and conflicts of interest, with fewer ethical constraints than his first term.

Accomplishments

- Signed the Tax Cuts and Jobs Act (2017), reducing corporate and individual tax rates.

- Appointed three Supreme Court justices (Gorsuch, Kavanaugh, Barrett), shifting the court's ideological balance.
- Brokered the Abraham Accords, normalizing relations between Israel and several Arab states.
- Launched Operation Warp Speed, accelerating COVID-19 vaccine development.
- Renegotiated trade deals, including the USMCA (replacing NAFTA) and agreements with China, Japan, and South Korea.
- Reduced federal regulations, cutting nearly 25,000 pages from the Federal Register.
- Defeated ISIS's territorial caliphate in Syria and Iraq.
- Implemented criminal justice reform through the First Step Act (2018).
- Moved the U.S. Embassy in Israel to Jerusalem, fulfilling a long-standing policy goal.
- Imposed tariffs on steel and aluminum imports, aiming to boost domestic manufacturing.

First Lady's Contributions

Melania Trump, serving as First Lady from 2017–2021 and again from 2025, focused on targeted initiatives:

- Launched the Be Best campaign (2018), promoting child well-being, online safety, and opioid awareness.
- Advocated for foster care and adoption reforms, supporting faith-based organizations.

- Oversaw White House restoration projects, preserving historical artifacts and decor.
- Represented the U.S. internationally, promoting education and women's empowerment on solo trips.
- Maintained a low-profile approach, emphasizing privacy and selective public engagements.

Personal Positive Traits

Trump's charisma and resilience have been central to his public image, enabling him to connect with supporters and withstand intense political and personal challenges. His business background and media presence reinforced his ability to project strength and influence.

- He was known for his charisma, media savvy, and resilience.
- His confidence and direct communication style, especially on social media, allowed him to connect with millions while bypassing traditional media.
- His business acumen, developed through decades of deal-making, translated into bold policy decisions.
- Supporters admired his tenacity, especially after surviving financial setbacks and political attacks.
- Following a 2024 assassination attempt, he leveraged the event to project strength and determination.
- His ability to galvanize a loyal base reshaped the Republican Party and gave voice to disenfranchised voters.

Personal Negative Traits

Despite his strengths, Trump's leadership and communication style also drew criticism and sparked controversy. His rhetoric, behavior, and legal challenges fueled sharp divisions in public opinion.

- His impulsiveness and polarizing rhetoric often alienated both critics and allies.
- His tendency to make false or misleading statements undermined trust in political institutions.
- His combative style, including disparaging minorities and political opponents, fueled accusations of divisiveness and authoritarian tendencies.
- His legal troubles, including a 2024 felony conviction for falsifying business records, raised ethical concerns.
- Critics argued that his focus on personal branding often overshadowed substantive policy.

Effects on Presidency

Trump's traits profoundly influenced the course of his presidency, shaping both his victories and controversies. His outsider appeal energized supporters, but his rhetoric and leadership style also deepened partisan divides at home and abroad.

- His charisma and outsider status fueled his victories in 2016 and 2024, appealing to voters frustrated with political elites.
- His approval ratings rarely exceeded 40%, reflecting the divisive nature of his leadership.

- He made aggressive use of executive power, issuing over 90 executive orders in his first month in office.
- These rapid actions achieved quick policy wins but also sparked numerous legal challenges.
- His combative style contributed to high staff turnover within his administration.
- His approach strained international alliances, with global approval of U.S. leadership dropping to 16% by 2020.
- His focus on economic nationalism and deregulation cemented a legacy that was transformative but highly controversial.

Pets

Unlike many presidential families, the Trumps stood out for their lack of White House pets, a reflection of Donald Trump's personal preferences.

- The Trump family did not keep pets in the White House during either his first or second term.
- Trump expressed disinterest in pets, once stating he did not have time for a dog.
- No records indicate that pets were kept during his second term as of 2025.
- This absence reflected his preference for a pet-free lifestyle.

Religious Persuasion

Trump's religious identity has often been debated, with his statements and actions drawing both skepticism and support.

- He identified as a Presbyterian during the 2016 campaign.
- In 2020, he claimed to be a nondenominational Christian.
- Critics noted his limited engagement with religious practices and questioned the depth of his faith.
- Some biographers suggested his use of religion was largely for political appeal.
- Supporters emphasized his defense of religious liberty, including protections for faith-based organizations.
- A 2017 survey found that 63% of Americans doubted his religiosity.
- Evangelical voters strongly supported him due to his alignment with their values on issues such as abortion and judicial appointments.

Interesting Anecdotes

- Trump dictated a 2015 health letter claiming he'd be the "healthiest" president, later revealed as exaggerated by his physician.
- His 1980s tabloid persona in New York, fueled by self-promotion, earned him the nickname "The Donald."

- At Wharton, Trump was known for skipping classes but acing exams, relying on charm and quick thinking.
- He purchased Mar-a-Lago in 1985, turning it into a private club after local opposition blocked his housing development plans.
- Trump's *The Apprentice* catchphrase, "You're fired!" became a cultural phenomenon, boosting his national profile.
- In 2025, he reordered White House portraits, placing a dramatic painting of himself post-assassination attempt opposite Obama's, emphasizing his defiance.

Ages at Death

As of June 20, 2025, Donald Trump (born June 14, 1946) is 79 years old and alive, serving as the 47th president. Melania Trump (born April 26, 1970) is 55 years old and also alive. No information on their deaths or burial locations exists, as they are living.

Causes of Death

Since both Donald and Melania Trump are alive as of 2025, no causes of death can be reported.

Burial Locations

With both Donald and Melania Trump alive, no burial locations are applicable.

Conclusion

Donald J. Trump's life is a testament to ambition, controversy, and reinvention. From a brash New York real estate tycoon to a transformative political figure, he redefined American leadership with a populist vision that both inspired and divided. His business successes, reality TV fame, and unorthodox presidency—marked by tax reforms, foreign policy shifts, and legal battles—cemented his legacy as a polarizing force. Melania Trump's reserved yet impactful tenure as First Lady complemented his style. While his personal traits fueled unprecedented loyalty and opposition, Trump's influence on politics endures, shaping debates about power, democracy, and national identity.

Addendum

The media has long influenced public perception, often with bias, but its recent hostility is unprecedented. For instance, critics and the media have questioned the president's and the first lady's Christian faith, despite their assertions. As a Christian, I believe critics need a clear standard by which to substantiate their claims. *Matt. 7:16* states, "By their fruits you will know them." *Micah 6:8* outlines God's expectations: to act justly, show mercy, and walk humbly with Him. However, not everyone holds to a biblical worldview, so these words will be interpreted differently.

It would be good to remember that the standard used to judge others will be the same standard used to judge oneself. While nothing prohibits one from evaluating situations or actions, it would be beneficial if harsh condemnation could be avoided and humility and self-reflection could be part of civil discourse.

To evaluate a president's effectiveness, one must consider the challenges the elected one faces, the context, and the era of his tenure. A nation should be thankful for a strong, talented, yet imperfect leader chosen to guide an imperfect people.

www.ingramcontent.com/pod-product-compliance
Lightning Source LLC
Chambersburg PA
CBHW050728010526
44107CB00009B/782